P9-DTM-774

Balancing Jobs and Family Life

Balancing Jobs

and Family Life

Do flexible work
schedules help?

Halcyone H. Bohen
Anamaria Viveros-Long

Temple University Press
Philadelphia

Temple University Press, Philadelphia 19122

Published 1981
Printed in the United States of America

Library of Congress Cataloging in Publication Data

Bohen, Halcyone H.   1937–
  Balancing jobs and family life.

(Family Impact Seminar series)
Bibliography: p.
Includes index.
1. Family—United States—Case studies. 2. Hours of labor, Flexible—United States. 3. United States—Officials and employees—Family relationships.
I. Viveros-Long, Anamaria, 1942–   joint author.
II. Title. III. George Washington University, Washington, D.C. Family Impact Seminar. Family Impact Seminar series.
HQ536.B63   306.8′7   80-25165
ISBN 0-87722-199-5

For Fred Bohen
and for Shawn, Kim, and Courtney

For John Long
and for Daniel and Andrea

The major reasons for our interest in this topic

Family Impact Seminar Series

The Family Impact Seminar Series is a series of books deriving from the research of the Family Impact Seminar, which is part of the Institute for Educational Leadership at George Washington University.

Balancing Jobs and Family Life: Do Flexible Work Schedules Help?
by Halcyone H. Bohen and Anamaria Viveros-Long

Foster Care and Families: Conflicting Values and Policies
by Ruth Hubbell

Teenage Pregnancy in a Family Context: Implications for Policy
edited by Theodora Ooms

Family Impact Seminar Members

WALTER ALLEN, Assistant Professor of Sociology, University of Michigan

NANCY AMIDEI, Director, Food Research and Action Center, Inc.

MARY JO BANE, Deputy Assistant Secretary for Program Planning and Budget Analysis, U.S. Department of Education

TERREL BELL, Commissioner, Utah System of Higher Education

URIE BRONFENBRENNER, Professor of Human Development and Family Studies, Cornell University

WILBUR COHEN, Professor of Public Affairs, University of Texas at Austin

BEVERLY CRABTREE, Dean, College of Home Economics, Oklahoma State University

WILLIAM DANIEL, JR., Professor of Pediatrics, University of Alabama School of Medicine, Birmingham

JOHN DEMOS, Professor of History, Brandeis University

PATRICIA FLEMING, Deputy Assistant Secretary for Legislation (Authorizations), U.S. Department of Education

ROBERT HILL, Director, Research Division, National Urban League

NICHOLAS HOBBS, Professor of Psychology Emeritus, Vanderbilt University

A. SIDNEY JOHNSON, III, Director, Family Impact Seminar

JEROME KAGAN, Professor of Psychology, Harvard University

SHEILA KAMERMAN, Associate Professor of Social Policy and Planning, Columbia University School of Social Work

ROSABETH MOSS KANTER, Professor of Sociology, Yale University

LUIS M. LAOSA, Senior Research Scientist, Educational Testing Service, Princeton

ROBERT LEIK, Director, Family Study Center, University of Minnesota

SALVADOR MINUCHIN, Professor of Child Psychiatry and Pediatrics, University of Pennsylvania

ROBERT MNOOKIN, Professor of Law, University of California at Berkeley
MARTHA PHILLIPS, Assistant Minority Counsel, Ways and Means Committee, U.S. House of Representatives
CHESTER PIERCE, Professor of Psychiatry and Education, Harvard University
ISABEL SAWHILL, Program Director, Employment and Labor Policy, The Urban Institute
CAROL STACK, Director, Center for the Study of Family and the State, Duke University

Preface

The Family Impact Seminar

There is a myth in our country that government is somehow neutral to families. In fact, all levels of government have policies and programs that affect families deeply.

Any family that has paid taxes, contributed to or received benefits from Social Security, married or divorced, benefitted from the G.I. bill, or been involved with public schools, foster care, welfare, child care, or the court system knows that government affects families. The Family Impact Seminar studies have identified 268 federal domestic assistance programs alone—administered by seventeen different departments and agencies—that have a potential of direct impact on American families. The question is not, Does government affect families? but rather, How does government affect families and how can policies that hurt families be repealed or reformed?

The Family Impact Seminar was created in 1976 to explore these kinds of questions. The Seminar is composed of twenty-four of the country's leading scholars and public policymakers concerned with families. They convene several times a year to provide leadership and guidance for the work of the Seminar's staff.

The Seminar is founded on a respect for the integrity, diversity, and privacy of American families; a conviction that government policies should strengthen families rather than weaken them; and a commitment to the idea that families themselves should participate in the decisions that affect them.

The original interest in family impact analysis—which subsequently led to the creation of the Family Impact Seminar—came from the 1973 Senate hearings on "American Families: Trends and Pressures," and the seminal statement of Walter Mondale, then chairman of the Senate Subcommittee on Children and Youth: "We must start by asking to what extent government policies are helping or hurting families." Many of those who testified at the hearings recommended that family impact statements be developed for all public policies. Although intrigued by the idea, the Subcommittee concluded that legislation would be premature before the new concept had been tested. I left my position as staff director of the Mondale Subcommittee to found the Family Impact Seminar in order to undertake this testing in an independent, non-governmental setting.

Family impact analysis, as we define it, is a process of assessing the effects of public or private policies on families. Its objective is to make policies and practices more sensitive to the needs and aspirations of families. It is designed to provide practical policy recommendations in a relatively short period of time.

The process of family impact analysis includes reviewing laws and regulations, interviewing policymakers and service providers, and learning directly from families how policies affect them. It lays considerable emphasis on assessing how programs really work. These are some examples of important family impact questions to ask about policies and programs:

- Does the policy encourage or discourage marital stability?
- Do families have real opportunities to participate in the decisions that affect them?
- Does the policy encourage caring for family members by families themselves or by institutions? Does it use or ignore existing family support systems, such as extended families and kinship ties?
- Is the program sensitive to the traditions, values, and prac-

tices of families from varied racial, ethnic, and religious backgrounds?

This volume addresses one of the most pressing problems confronting American families—how employed people can balance their job and family responsibilities. The Seminar's other two volumes analyze the effect of foster care policies and teenage pregnancy policies on families.

These case studies are the first comprehensive efforts to test the value of our evolving framework for family impact analysis. They are our attempt to take some promising concepts, apply them to existing public policies, and discover if they yield useful findings and recommendations.

The results of this work have far exceeded our expectations. The talents, insights, and dedication of each study director—backed up by generous support from interested funding sources—transformed the projects from relatively short-term policy reviews to long-term, in-depth studies. Each study uses a different design and methodology: survey questionnaires and group interviews in the work schedules study; interviews at the service delivery level in the foster care study; and commissioned papers and a conference in the teenage pregnancy study. But these studies have fundamental elements in common, including a focus on the diversity of families, an appreciation of the different levels of policy implementation, an emphasis on obtaining the views of the families being served, and a reliance on a family impact perspective and framework.

In addition to the valuable advice received from Seminar members and other experts, this volume primarily reflects the complementary talents of its two authors. Halcyone H. Bohen directed the study from its inception, coordinating all its elements over three years. She collected the background government data, reviewed the literature, conducted the interviews, and wrote the manuscript. Methodologist Anamaria Viveros-Long led the planning of the survey research design and instrument development and managed the analysis of the statistical data. The insights and analyses of the book reflect both authors' professional expertise, as well as their experiences as spouses and parents. Bohen is a psychologist, trained at New York University and the Philadelphia Child Guidance Clinic, and a former Princeton University dean. She is presently investigating the effects of U.S. and European corporations on employees' families for the Children's Defense Fund, the Foundation for Child Development, and the German Marshall Fund. Viveros-Long is a sociologist trained at the University of North Carolina. She spent a year on this study after working on several research projects in Chile and the U.S. Office of Education. She is now a social science analyst for the Office of Evaluation at the U.S. Agency for International Development.

Our case studies provide three different promising models of family impact analysis for organizations with a national perspective and considerable resources of time, funds, and expertise. Based on the case study experiences, we have also developed a family impact approach for individuals and organizations who work with families on a daily basis, but have limited time and resources to devote to policy analysis. With the help of twelve organizations across the country—four PTAs, four antipoverty agencies, three children's agencies, and one hospital—we are currently exploring whether family impact analysis is also a useful tool for state and local agencies. Policymakers, service providers, and families themselves are being interviewed; laws, regulations, guidelines, and memoranda are being dissected and analyzed. While this field project is not fully completed, we are encouraged by its progress to date. We are already gaining insights into how policies are affecting families in different communities, as well as ideas about how policies may be reshaped in order to be more helpful to them.

Based on our work over the last four years in case studies and field projects, the Seminar recommends several steps to make public and private policies more responsive to the needs and aspirations of families. We propose that: (1) independent Commissions for Families be created at several levels of government to conduct family impact studies, and help ensure that public policies aid rather than hurt families; (2) public and private organizations and agencies examine and improve the ways in which their own policies and practices affect families; and (3) more organizations of families themselves assess and improve the impact of policies or programs on families.

Although these suggestions alone will not solve all the problems that face families, they can be a first step toward the day when decision makers consciously and consistently ask, "What effect will this policy have on families?"

A. Sidney Johnson, III
Director, Family Impact Seminar

Contents

Tables and Figures

Tables

Figures

Acknowledgments

Our colleagues at the Family Impact Seminar were the heart of this project for three years. Director Sidney Johnson's confidence, perspective, and kindness characterized his role in our work, as in the other imaginative activities he has conceived and launched. With unstinting respect for the complexity of family impact analysis, Deputy Director Theodora Ooms kept easy answers to our inquiry at bay and appreciation for our efforts paramount. Ruth Hubbell, Seminar research director, shared her technical expertise and provided cogent critiques of our drafts. With sensitivity to personal and work pressures, Susan Farkas was an ideal officemate for the last year of the project.

Mary Eng's dedication and skill in both technical and human dimensions of the project were supreme contributions to its completion. Her library research, typing, keypunching, and preparation of tables and references were flawless. Darlene Craddock's cheerful readiness to help with any need kept us bright on bleakest days. Elizabeth Bode's delight in language and her exacting copy editing buoyed our spirits throughout the manuscript preparation. John Sheldon handled the project finances with assiduous attention to detail and timetables. Yvonne Stoltzfus and Edward Metz shared secretarial responsibilities at the beginning and end of the study.

The members of the Seminar who served as the Advisory Committee for this project—Mary Jo Bane, Urie Bronfenbrenner, Beverly Crabtree, John Demos, Sheila Kamerman, Jerome Kagan, Rosabeth Kanter, Robert Leik, and Isabel Sawhill—all gave generously of their time and interest. Two Seminar members were especially important to our effort. In addition to his theoretical contributions, Urie Bronfenbrenner's dauntless enthusiasm for this project, for the causes it represents, and his persistent questions of the data were steady sources of inspiration and support. The formulation of key questions about work and family issues by Rosabeth Kanter provided major organizing principles for the research; her files on the topic, her quick insights and energy provided sustaining momentum.

Several non-Seminar members of the Advisory Committee also contributed countless hours to thinking through the research design, the measurement questions, and the interpretation of the data. Foremost, for three years Mary Sue Richardson of New York University provided tough-minded, critical, and benign guidance. Joseph Pleck of Wellesley College helped frame the basic analysis and shared abundantly of his own research and time. Sharon Weinberg of New York University solved numerous statistical conundrums with ease and clarity.

Donna Dempster of Cornell University and California State University at Long Beach reviewed the literature on participant observation, helped to plan and conduct the interviews, to transcribe the video tapes, and to interpret the findings—and contributed to countless other Seminar activities in the summer of 1979. Dorothy Ross of The University of Virginia gracefully clarified the historical themes in Chapter 2.

Beyond the financial support of their organizations, we also benefitted from the keen attention of our foundation program officers, Barbara Finberg and Gloria Primm Brown of the Carnegie Corporation, and Jane Dustan of the Foundation for Child Development, who read our materials, offered ideas, and attended our meetings with sustained involvement. Elliot Liebow, chief of the Center for Studies of Metropolitan Problems at the National Institutes of Mental Health, arranged funding to support part of the research and alerted us to nuances of our topic with humor and warmth.

Consonant with her long-time efforts on behalf of employment opportunities for women, Elsa Porter, assistant secretary for administration for the Department of Commerce, and her staff opened doors for the research in their department. Lorin Goodrich, deputy director of the Office of Management and Administration of the Economic Development Administration, Peter Hannums of the Office of Personnel of the Maritime Administration, and Marilyn McClennan of the Office of Organization and Management System of the Department of Commerce genially made arrangements for our data collection in the agencies. Barbara Fiss and Thomas Cowley of the Pay and Leave Section of the Office of Personnel Management taught us the basic information on federal flexitime, and Cowley patiently supplied answers to many technical questions. From her own research on hours of work, Virginia Martin of Vail Associates sharpened our inquiry with important contributions to instrument development and data presentation. Linda Ittner, staff assistant to Congresswoman Patricia Schroeder, kept us abreast of alternate work schedule legislation through 1977 and 1978.

At the survey data collection and coding stages we were ably assisted by the questionnaire distributors, Edgar Boling, Carol Doolan, Ellen Mulroney, Delores Royston, and Randy Swisher; and by coders Roselyn Dixon, Bonnie Oglenski, and Gary Peck. Jane Scott eagerly served in both capacities. At the interview stage, Donald Budowski of the Office of Organization and Management System of the Department of Commerce guided us through the Privacy Act protections.

John Long of the Bureau of the Census provided expert answers to numerous methodological and demographic questions. Michael Frodyma of Mesa Data Services efficiently programmed the survey data. Chaya Piotrkowski of Yale University offered perceptive suggestions on our first draft. We also benefitted from several useful conversations with Sandra Hofferth and Ralph Smith of The Urban Institute. Heather Weiss, director of Data Analysis for the Cornell Family Stress and Support Project, talked with us about conceptual and methodological challenges of "ecological" research. Allyson Grossman of the Bureau of Labor Statistics had ready employment data whenever we called. With short-term dispatch and accuracy, Yoma

Ullman of Princeton, New Jersey, edited the manuscript before submission to the publisher. Mary Procter generously loaned her camera equipment for the photographs and provided humorous and pithy tutelage on the pleasures of its use. The sponsorship of Temple editor Mike Ames resulted in this published account of the research. With unshakable calm, Merry Post of Temple University Press coolly guided the manuscript to punctual publication.

Finally, we gratefully acknowledge funding from the Carnegie Corporation of New York, and the Foundation for Child Development, who provided the primary financial support for this study. The Center for Studies of Metropolitan Problems of the National Institutes of Mental Health helped support the interview phase in particular. We also wish to thank the following for their contributions to the study and to the Seminar: Lilly Endowment, Inc., Ford Foundation, Mary Reynolds Babcock Foundation, Robert Sterling Clark Foundation, Needmor Fund, General Mills Foundation, Levi Strauss Foundation, Chichester duPont Foundation, Administration for Children, Youth, and Families, Community Services Administration, Edna McConnell Clark Foundation, and Charles Stewart Mott Foundation.

Halcy Bohen also especially thanks James and Eone Harger, Penelope Passage, Benita Eisler, and Angelica Rudenstine for essential support in ways they each know.

Balancing Jobs and Family Life

1

Introduction

FOR 95 percent of employed Americans, the places where they live and the places where they work are two separate worlds. Typically a half-hour's travel time lies between job and home. Yet people have daily responsibilities in both settings, especially if they have children who depend upon their care.

Most of the daylight hours, most days of the week and year, the children are one place and their parents are somewhere else. Both parents of a majority of American children are employed, including almost all the fathers and half of the mothers. Depending upon their ages, children are at school, at home, in child care facilities—or in a variety of other places doing things with or without adults other than their parents.

These characteristic circumstances of work and family life in the United States in the late 1970s led to interest in the topic investigated in this study. Would giving employed adults some flexibility in scheduling their hours of work help them to function successfully in both their job and family worlds, for their own sakes and those of their children?

Chapters 1 to 3 explain how the question was addressed in a survey of seven hundred federal government employees, in light of historical and contemporary perspectives on the scheduling of work. Chapters 4 to 8 explain how we designed and executed the study and what our findings were.

In Chapter 9 and the Afterword, we draw general conclusions about our work-family topic and evaluate the concept of family impact analysis and our research in terms of our experience with this project. The results reflect the complicated and shifting questions of cultural values about men and women's roles in work and family that lie behind the seemingly simple central question of the study. The entire project represents an attempt to analyze the impact of a public policy on American families.

The origins of the present study developed in the early 1976 discussions by the Family Impact Seminar of the effects of government work policies on families. Chapter 1 describes the steps taken to narrow the Seminar's interest in the subject to a feasible one-year case study for family impact analysis. The following account of the way the project was approached is provided for two reasons. First, it indicates the range of issues encompassed by the relationship of work policies to family life, as well as the particular aspects of federal employment that were reviewed. Second, it illustrates the kind of "focusing" process that any group choosing a broad topic for family impact analysis may need to undertake.

The account also suggests two ways to define *family impact analysis*. On the one hand, it may be thought of as the entire saga of meetings, telephone calls, literature reviews, consultations, and decisions through which the focus on flexitime policies was established—plus the survey design, data collection, analysis, and follow-up interviews. On the other hand, it may be exclusively thought of as the methodology by which we attempted to measure the effects on families of one particular employment policy in one federal department. This chapter reflects the more comprehensive definition of the process by providing a step-by-step account of the preliminary activities undertaken to arrive at a topic.

When the Family Impact Seminar initially considered how to analyze the impact of public policies on families, the members briefly reviewed ten topics for possible attention: teenage pregnancy, tax policies with regard to families in which both husband and wife are employed, incentives for institutional care of the aged, welfare, housing, agricultural extension programs, unemployment, government as an employer, foster care, and mental health programs. From this list the Seminar chose the topic of government-as-employer for its first experiment for the following reasons:

- The federal government is America's largest employer. Its employment practices, therefore, affect large numbers of people. Of a total United States population of 217 million, five million people work for the federal government worldwide.
- The policies of the federal government provide models for other levels of government and for the private sector. Even when the government is not the first to develop a program, replicating it in the federal system gives the practice wide visibility and a stamp of approval that makes it seem "safe" for other employers.
- On-going experiments with federal employment policies and pending bills to expand these efforts contributed to the feeling among Seminar members that the late 1970s was an auspicious moment to examine the impact of federal employment on families—the momentum of change was already evident.

On the basis of the above rationale, and to provide a comprehensive basis for selecting a particular policy for impact analysis, we listed as specifically as possible the categories of employment policies that might affect the family lives of employees. They are:

- Availability of jobs. How do fiscal and monetary policies affect the existence and location of jobs?
- Eligibility criteria. What are the requirements for obtaining or advancing in federal jobs (e.g., age, sex, training, and experience)?

- Compensation and benefits. What are the pay scales and fringe benefits? What are health, leave, vacation, and retirement policies?
- Location and transportation. Where are the places of employment in relation to residential areas? What transportation options are available in terms of means, schedules, cost, and comfort?
- Quality of work life. What is the environment of the job—intellectually, psychologically, socially, and physically?
- Job security. How certain is it? What is it based on?
- Hours of work. How much choice is available for four-day weeks, part-time, overtime, shifts, flexitime, flexiplace, and the like?
- Child care. Is on-site child care available, and/or is there neighborhood child care?
- Transfers. How often is relocation required and with what choice and/or support services?
- Expectations for spouses. Is there an implicit, or explicit, policy that spouses are available to contribute to the work and advancement of the employee by entertaining, child rearing, and so forth?
- Availability of consumer services. How convenient to places of work are markets, post offices, and recreation?

With this overview of particular employment policies, we established four criteria for narrowing our focus for a topic for family impact analysis:

- The policy should affect large numbers of people.
- The available data should enable us to demonstrate the suitability of the policy for family impact analysis.
- The issue should have enough visibility in the policy-making process to bring attention to the family impact analysis approach.
- The policy should give families the support or opportunity to deal with their needs in their own ways.

With these criteria in mind, we narrowed the above list of eleven employment policies to four candidates: child care, quality of work life, transfer policies, and hours of work. Each of these was considered in the following way.

Child Care

Child care met each of the four criteria for selection in several respects. First, the availability of quality child care unquestionably has an impact on large numbers of families by affecting the well-being of children and the time parents of young children have for employment. In addition to whether jobs are available, child care may be the major determinant in whether mothers are employed, given the fact that they continue to have primary responsibility for child rearing in this country. Women affected by the availability, cost, and quality of child care services include not only the 34 percent of mothers with children under age six who were already in the labor force in 1977, but also those who would wish to be employed if they had good child care options, as well as those with school age children who need after school care.

Second, the Women's Bureau in the Labor Department had complete and up-to-date information. In 1977, the federal government had nine on-site child care centers in the Washington, D.C., area. All the centers were of high quality and in great demand, but they had been created largely through the volunteer efforts of employees. At that date, the centers served a total of 388 children while in the Washington, D.C., area the federal work force (civilian and military) totalled approximately 500,000; perhaps one-twentieth of these workers (a conservative estimate) had children under ten years of age who needed care apart from already existing public school facilities. In short, the available centers probably served only a fraction of the eligible children of federal employees.

Third and fourth, the issue of child care had already achieved national visibility, both before and after Richard Nixon's 1971 veto of child development legislation that would have increased federal money for day care; but the topic was still a subject of complex and controversial national debate. One level of discussion concerned whether or not the government should encourage out-of-home care for young children at all. Disagreements revolved around ideas about both the best ways to rear children and the appropriate social roles of women. On the second level, child development specialists, as well as parents and public officials, debated what kind of out-of-home care is best for children. For example, while few persons advocated expanding "custodial" care for children, many spokespersons agreed that small scale community-based "family day care" is preferable to child care in workplaces. On another level, controversy continued about what role, if any, employers should have in providing or helping support any kind of child care for employees' children. Ultimately the Seminar decided that this subject, like welfare, was too complex and controversial for the Seminar to address in an initial experiment with family impact analysis.

Quality of Work Life

Among the employment issues considered, the meaning of *quality of work life* (QWL) as a "policy" suited to family impact analysis was the most difficult to specify. This definitional difficulty effectively disqualified the topic as a candidate for the experiment in family impact analysis. In the last decade, quality of work life has become a shorthand term for certain work-reform groups to refer to a process in which employees have a regular and effective voice in determining the nature of their working environment, their work relationships, their work content, and their compensation. One organization has explained that QWL is basically "the Boy Scout oath . . . applied to work organizations." (See further discussion of these reform groups in Chapter 4.)

While most Family Impact Seminar members believe that the quality of work life inevitably affects workers' personal and family lives, the topic seemed too vague for family impact analysis. Although the Seminar was aware of the efforts of individual managers from several agencies (including the World Bank, ACTION, the Postal Service, and the Commerce Department) to introduce QWL reforms in their agencies, these activities seemed too few, too scattered, and too amorphous to investigate systematically. Therefore, the Seminar moved on to another topic.

Transfers

Transfer policies met several of the selection criteria. The families of military and foreign service personnel (plus those of other agencies with overseas operations—the Central Intelligence Agency, United States Information Agency, Agency for International Development, and the Agriculture and Commerce Departments) are regularly affected by the procedures of rotating personnel among different posts. Many companies in the private sector have similar policies, usually intended to broaden executives' experiences as part of their career development. In the federal government this employment practice potentially affects approximately three million Department of Defense employees (two million uniformed and one million civilian personnel) and nine thousand foreign service personnel. Civil service employees in non-military, domestic positions are not required to be willing to relocate as a condition of employment; and therefore the Civil Service does not have a transfer policy per se. Yet sections within a federal agency occasionally are shifted in their entirety from one locale to another.

The complications created for families by transfers are similar in both the military and the foreign service, but the family-related policies of the two services have varied. For example, the military operates schools for the children of its personnel on overseas bases, whereas the Foreign Service does not. The spouses of both foreign service and military personnel have traditionally been considered extensions of their spouses, in the sense of representing United States interests in the communities and countries where they were stationed, although they have not been directly hired or compensated for these roles.

In the 1950s, the military created a range of support services to ease the effects of transfers on families. Although the programs of the Army, Air Force, Navy, and Marines vary, all of them provide medical care for dependents, commissaries, post exchanges, educational loans, legal services, moving allowances, recreational facilities, and schools.

While information on service transfer policies is available, it is difficult for groups like the Family Impact Seminar to find out how many people are affected annually. The army officer who eventually tracked down the information for us (at the behest of a congressional referral) had to make almost a dozen calls himself to compile the information. He finally determined that in fiscal year 1976 the Defense Department moved almost one million dependents of uniformed married personnel. Although transfer policies were considered good candidates for our family impact analysis for the above reasons, hours-of-work policies were ultimately chosen because they even more appropriately met selection criteria, as discussed below.

Hours of Work

Like quality of work life, regulations about hours of work affect all federal employees. But unlike specific QWL policies or activities, which had been introduced in only a handful of offices in the federal government by 1977, hours-of-work policies have always affected every federal employee and several kinds of policies could therefore be examined for their effects, including traditional standard schedules, overtime practices, shift work, part-time policies, and flexible scheduling options.

Approximately one hundred forty-one thousand federal employees had the option of flexible work schedules by the fall of 1977. Pending legislation would experiment with even more flexibility for many more thousands. As the bills were debated, the Government Accounting Office, a research arm of the Congress, compiled information on existing federal experience with scheduling options (U.S. Comptroller General 1977). The bills awaiting a congressional vote required additional evaluations of any future programs. A number of private sector firms also had conducted evaluations of their flexible scheduling experience.

In addition to the potential of hours-of-work policies to affect all levels of federal employees, the spate of popular articles on alternate work scheduling, and the favorable testimony in congressional hearings on the pending bills, gave the topic relatively high visibility among the issues in the policymaking process the Seminar considered. Finally, as a policy that simply gave workers more choice about when to be at their offices, without containing explicit values about family life, hours-of-work policies met the criterion of giving families opportunities to deal with their needs in their own ways.

At this juncture we decided we needed answers to several additional questions before designing a process to assess the family impact of any of these hours policies:

- What data were available on the families of federal employees?
- What were the existing federal policies on hours of work regarding part-time, overtime, and flexible scheduling?
- Had any studies, reports, or evaluation of alternate work schedules considered their effects on families?

Family Data on Federal Employees

Through dozens of telephone calls to various government records offices, we learned that the federal government has virtually no reliable aggregate data on the family characteristics of its employees. The government does gather some family information on its employees, but in several different ways and for a variety of purposes, none of which added up to useful information for our family impact analysis. For example, for the purpose of health insurance, close to 95 percent of federal employees submit information on their age, sex, marital status, and the ages and names of their dependents. But this information is kept in thousands of separate federal payroll offices, filed only by employee name and number. The health insurance carriers themselves file the family information only according to whether the contract is for an individual or a family.

Federal employees also supply family information on income tax deduction statements, but the data on such forms is often incomplete, depending on the rate at which a taxpayer wants taxes withheld (e.g., more or fewer dependents may be claimed). The family data on federal job applications—the Standard Form 171 in use since 1940—appear only in terms of how the applicant wishes to be addressed (Mr., Mrs., Miss) and serve mainly to provide statistics on hiring practices (for example, the number of applicants of each sex compared with the number of hires). Whatever aspect of hours-of-work policy was to be analyzed for its impact on families of federal employees, it appeared that we would have to collect the relevant family data ourselves.

Federal Hours-of-work Policies

Changes in work time for the federal civil service have reflected shifts in the private sector. As recently as 1900, ten-hour days and six-day weeks were the norm for American workers. The Fair Labor Standards Act of 1930 established the *standard week* in which premium wages were required for hours worked in excess of forty per week; this law affected employees engaged in interstate commerce but not employees of the federal government. As late as 1938, 60 percent of federal employees worked more than five days a week. It was not until 1945, with the Federal Employees Pay Act, that overtime pay was required for federal employees who worked more than forty hours a week. And the provision (of Title 5 of the United States Code) requiring federal employees to work eight-hour days was not enacted until 1954 (Civil Service Commission 1962).

Most federal civilian employees today officially work eight-hour days. In practice there are variations on the high and low ends. Lateness, long lunches, or other ways of skimping on the prescribed forty hours per week are difficult to counter except through the administrative rigor of particular supervisors. Dismissal of a civil service worker requires elaborate and lengthy steps; the compexity of the process is such that federal employees are seldom fired.

Federal employees at the highest levels routinely work more than forty hours a week—to handle heavy workloads, to nurture promotions to higher levels of responsibility and pay, or for other psychological reasons (to satisfy needs for achievement and recognition, to enjoy the challenge of their assignments, or to avoid doing other things, including in some cases participation in family life). Although they do not have nearly the freedom of choice of private sector professionals (academics, self-employed physicians, attorneys, and the like), the higher their civil service grade, the more authority government workers have to make their own choices about whether to do non-job activities during normal working hours; for example, to come late, to take off time for appointments, or to leave early. The success of the job is measured by what is accomplished, and to some extent by total hours logged in, but not necessarily by the regularity of the hours. As in the private sector, "production" units tend to be the exception to this general proposition; that is, those government offices that must keep up with a daily output, or service to the Congress or public, require their executives and professional personnel to be on the job as regularly as the non-professionals.

Part-time Employment in the Federal Service

The term *part-time* is most often used to mean working, by choice, for fewer hours per day or week than is typical in a given location. But it is difficult to distinguish between those working part-time voluntarily and those who wish full-time work (International Labor Organization 1973).

For federal government employment, the Standard Form 171 permits applicants to indicate their willingness to work part-time. However, no roster is kept of those interested in this option.

In terms of the types of work available on a part-time basis, federal employment mirrors the private sector, in which the bulk of the part-time work takes place in the lower levels of food service, clerical work, retail sales, and cleaning positions. The number of part-time workers has nearly doubled in the last twenty years in the United States as a result of sharp increases in the number of those seeking work in the groups traditionally interested in part-time jobs, namely, women, students, and older workers (Eyde 1975; Owen 1975).

In the federal service, few agencies use part-time workers in professional or semi-professional jobs, but almost none of these work in supervisory or management positions. At the time the Seminar was reviewing policies, the "ceilings" policy operated as a brake on the expansion of part-time opportunities in the federal government. This policy gave agencies an upper limit, or ceiling, on the number of employees it could hire—and part-timers were counted like full-timers, that is, as occupying one slot each. However, this stipulation was changed by the law to expand part-time opportunities that was passed in October of 1978. Beginning in the fall of 1980, total hours, instead of persons, will be counted for each agency. Since the 1977 presidential message encouraging increases in part-time hiring—and the 1978 legislation—twenty thousand new part-time positions have been created in the federal government; but only 30 percent are above grade 5 (McHugh 1980).

Overtime in the Federal Government

Information on hours worked overtime in the federal government is not compiled for individuals who work overtime except in individual payroll offices. Instead, the total number of hours worked overtime is counted to calculate the additional cost for paying overtime workers at one and one-half times their regular hourly wage.

The Civil Service Commission does compute the average figure of overtime worked per federal employee per week, which is 1.13 hours. Thus, the "average" federal employee works very little overtime—only twelve minutes a day. But this figure is not very useful in assessing federal service overtime since most overtime occurs in production units and/or with high-level professionals. Volunteer overtime technically does not exist since supervisors must request and receive approval for overtime in advance.

Generally, employees who are exempt from the provisions of the Fair Labor Standards Act (those at service grades 10 and above, who constitute 40 to 50 percent of the federal labor force) seldom receive overtime pay, although they are eligible. That is, a mid- or senior-grade employee could work overtime and get paid up to a total annual income of $47,025 (as of 1979). More often, they take *compensatory time* in exchange for longer hours worked (which is given at the rate of one hour off for each extra hour worked, not, as with money, at time and a half).

High-level professionals, however, are routinely expected to work more than forty hours per week to fulfill their assignments; this expectation is not spelled out in any civil service regulations but is an unwritten *sine qua non* of promotional opportunities in the higher grades. They seldom collect their "comp time." Employees in this category may constitute 5 to 10 percent of the federal work force.

Federal Government Flexitime

Flexitime means essentially that workers have some choice about when to begin and end work each day. Each workplace sets up its own system, but usually there are both flexible bands—in the morning, mid-day, and afternoon—and "core" times—usually during mid-morning and mid-afternoon—when everyone must be present.

Flexitime was officially introduced in the federal government in 1974, when a manager in the Social Security Administration headquarters in Baltimore requested that the Civil Service Commission grant a waiver from Title 5 of the United States Code, which reads, "working hours must be the same each day." At that point the Civil Service Commission determined that because the law was intended to protect employees from arbitrary designation of work hours by employers, "working hours" could be interpreted to include all the hours during the day when an employee is permitted to work. Thus, the Baltimore experiment proceeded in accordance with the law; when initial evaluations showed improved morale and reduction in the use of sick time and overtime, the program was systematically expanded throughout the Social Security Administration. It was learned subsequently that as early as 1972 the Bureau of Indian Affairs (BIA) had developed essentially the same system on its own; thus, the BIA probably had the first federal flexitime program, but unofficially. By the fall of 1978, about two hundred thousand federal employees were on flexitime, that is, about 10 percent of the federal government civilian workforce.

In 1977, when we began the study, all federal flexitime programs required employees to work eight hours a day—no less and no more (except at the employer's request and at overtime pay). Thus, the version of flexitime being considered in this study allowed people simply to start or end work a couple of hours earlier or later each day. At that time most federal flexitime arrangements were *modified*, that is, employees could pick their daily starting and ending times, but had to keep to those times, with a fifteen- to thirty-minute leeway period, unless they arranged in advance for exceptions. Thus, flexitime in the federal civil service meant that people worked from whenever they started in the morning until they reached eight and a half hours a day—unless they "flexed" at mid-day for longer than the thirty-minute lunch break.

Evaluations of government flexitime, like those in the private sector, reported high levels of satisfaction from employers and employees. In 1977, the General Accounting Office (GAO) surveyed thirty government organizations on flexible schedules (representing 133,000 of the 200,000 employees on flexitime). Favorable effects most often mentioned included: less tardiness (83 percent); less absenteeism, less short-term leave taken (71 percent); longer service hours for the public; and higher employee morale (86 percent) due to more job satisfaction, easier commuting, easier child care arrangements, and control of one's own time. "Improving employee morale" was the most common objective in introducing flexitime. According to the GAO, the organizations it surveyed felt that morale improved when people were "allowed more freedom to control the work situation and to assume responsibility for their own actions" (U.S. Comptroller General 1977, p. 11; Cowley 1976).

While the Seminar was considering which work policy to select for family impact analysis, legislation was pending that would expand flexitime and part-time options to approximately five hundred thousand more federal workers in the course of three-year experiments. These new federal experimental programs also significantly expanded the concept of flexitime. The law authorized agencies to let workers choose to work either more or less than eight hours each day, as long as they averaged eighty hours each fortnight (an arrangement called *banking and borrowing of hours*). The law also authorized experiments with compressed schedules (e.g., four-day weeks). The Office of Personnel Management (formerly the Civil Service Commission) was to monitor and evaluate the impact of the experiments on (1) the efficiency of government operations; (2) mass transit facilities; (3) levels of energy consumption; (4) service to the public; (5) increased opportunities for full-time and part-time employment; and (6) families and individuals generally (U.S. Congress, HR 7814; Causey 1976, 1978).

The legislation passed and was signed by President Carter in September 1978. The mandate to evaluate the effects of families of increased use of flexible work schedules resulted from an amendment to the House of Representatives bill proposed by the Family Impact Seminar, in response to a request from the Committee on Post Office and Civil Service that the Seminar comment on the pending legislation.

Existing Evaluations of Flexitime

Although both public and private sector self-evaluations of flexitime had been very favorable, little independent empirical research on flexitime had been completed before 1977. The few existing studies of flexible work hours were based largely on employer records and attitude surveys. Moreover, almost all the research and evaluations focused exclusively on workplace factors—attitudes, productivity, reductions in absenteeism and tardiness, and the like.[1]

Most important for our purposes, no systematic comparison of flexitime workers with those on standard work schedules had investigated the assumption that flexitime would be beneficial to a worker's family life. Of the dozen agency evaluations that the Office of Personnel Management had collected by 1977, only two asked questions directly related to family issues. Three studies of flexitime in relation to family issues were in process. Winett and Neale (1978) collected several kinds of family-related data from two small groups of federal employees, including time-diaries in which respondents reported on their use of time

[1]See review article by Nollen 1979; Social Security Administration 1974; Schein, Maurer, and Novak 1977; Golembiewski et al. 1974; Golembiewski et al. 1975; Owen 1975; Martin 1975; Hedges 1977; Kuhne and Blair 1978; Silverstein and Srb 1979.

in the weeks preceding and following implementation of flexitime in their agencies. Allan R. Cohen (1978) gathered questionnaire data from employees of the John Hancock Company in Boston that indicated that workers increased their family time and activities after they started working on flexitime. Portner (1978) released preliminary results on a study of people who elected to use flexitime in a Minnesota corporation. All three reports were in progress.

Once we had assembled information on the three main issues—that is, what data on federal workers were available, what existing federal hours-of-work policies were, and what reports and studies on alternate work schedules existed—the Seminar narrowed its choices for a first experiment in family analysis to flexible work schedules, for the following reasons.

Seminar members believed that a family impact analysis of flexitime would provide an opportunity to test the method itself in terms of a systematic policy available to all levels of employees, as opposed to a policy that had more categorical constituencies (like child care or transfer policies, for example). Second, the members thought that flexitime would have advantages for families of different structures and at various life cycle stages. Finally, with the selection of flexitime, the Seminar reaffirmed its special concern for families with children and its intention to focus the analysis on them.

Once the Seminar had resolved to analyze the effects of flexible work schedules on federal employees, we proceeded by posing three more questions to help us address the problem:

1. Why has interest in flexible work hours arisen now? What has changed in the relationship between work and families?
2. Who thinks flexitime will help families, and how?
3. How have work and family questions been examined heretofore in both scholarly and public policy contexts?

The next three chapters present the answers we developed.

2

Changes in the Relationship between Work and Families

IDENTIFYING the aspects of family life that people now hope to alter by means of flexible work schedules requires some understanding of the characteristics of work-family relationships in earlier eras. What is it people wish to restore, or to create anew, in work-family connections by this change? What values persist and what diverge from earlier experiences?

In the last decade, a number of social historians have posed research questions that can lead to answers to these questions. In contrast to the emphases of earlier historians on public events and elite groups, contemporary scholars are attempting to understand the broader structures of everyday lives in the past. They try to depict the social conditions that have affected people's life choices in particular settings at various life stages. They have constructed demographic, economic, and cultural portraits of social conditions by using heretofore untapped institutional records—such as census, church, business, court, and housing documents—as well as personal writings—such as diaries and letters—and popular artifacts like sermons, manuals, calendars, magazines, and engravings.

Recently a few of these "new" social historians have begun to examine the connections between work and family structures in the past. Their reports suggest that the patterns of work-family relationships among people living in the same communities at the same time—to say nothing of larger differences between countries and eras—have varied widely. As more information is assembled, it is increasingly clear that comparisons about family life across time, place, class, and sex must be tentative. With this cautionary note about the hazards of generalization on this topic in mind, in Chapter 2 we offer a brief account of earlier work-family circumstances as a context for our focus on work schedules in this study.[1]

Three fundamental changes in work-family relationships in Europe and North America, over time, have created the basis for the current interest in flexible work scheduling:

1. The separation of workplace and home life.
2. The drop in the birth rate since the eighteenth century.
3. The increase in labor force participation by women with children.

The present day concatenation of "problems" resulting from these historical changes has been characterized as follows:

> Family scheduling problems have increased dramatically. Working couples with school-age children probably have more schedule conflicts than did the average family in 1900. The fragile and harried lifestyles of families with two working parents is dramatically illustrated whenever there is an emergency closing of schools, as occurred in many communities during the energy shortage [in] . . . January 1977. Backup facilities

[1]The following were consulted in developing the ideas in the chapter: Ariès 1962; Bane 1976; Bennett and Elder 1979; Bernard 1974; Chafe 1976; deMausse 1974; Demos 1974; Greven 1970; Gronseth 1978; Grossman 1979; Gutman 1976; Hareven 1975, 1976; Hareven and Langenbach 1978; Hoffman and Nye 1974; Hunt 1970; Illick 1974; Barbara Laslett 1973; Peter Laslett 1977; McLaughlin 1971; Moore and O'Connell 1978; Myrdal and Klein 1956; Elizabeth Pleck 1976; Rosenberg 1975; Rothman 1978; Scott and Tilly 1975; Smuts 1959; Stone 1975; Taeuber and Sweet 1976; Tucker 1974; Waldman et al. 1979.

> for massive child care needs simply do not exist, which caused many family "crises" and much absenteeism at work.
>
> Haldi 1977, p. 233

For the purposes of this study, the evolution of these developments can be seen in terms of three stages in Western family history. These stages did not occur simultaneously across all American and European settings, but the sequence of change in family-work relationships remained the same for different places and classes. The relevant stages are:

Stage One. When workplace and home were one and the same.
Stage Two. When men's work took place away from home.
Stage Three. When women with children worked away from home.

Stage One: When Workplace and Home Were Essentially the Same

In various forms among different classes and occupational groups, this pattern of home as a workplace seems to have prevailed before industrialization (or modernization) whenever and wherever the latter occurred—in England, Europe, and the United States in the seventeenth and eighteenth centuries—and in other places more recently.

Evidence for the integrated nature of family-life and work-life settings before the eighteenth century is scanty, and historians differ in their evaluations of the beneficial qualities of the experience when it did occur. While there seems to be a consensus that the concept of childhood did not exist in pre–seventeenth-century Europe and America, one group of historians considers the quality of children's lives in pre-industrial times to be appalling, while another group finds many favorable features. Both interpretations are drawn from evidence of the highly integrated nature of work and family living.

Those who view pre–eighteenth-century family life with dismay cite evidence of the neglect and abuse of children, including high infant mortality rates, poor hygiene and nutrition, swaddling, leaving infants alone for lengthy periods; also they point to noisiness, crowdedness, lack of privacy, and the unsanitary conditions of housing, as well as routine wet-nursing, apprenticing of young children away from their families, and frequent brutality and a general lack of concern for children's vulnerabilities (deMausse 1974; Tucker 1974; Illick 1974).

On the other hand, those who see desirable features in the older lifestyles, point approvingly to the regular presence of adults and children in the same settings, as a result of the fact that work and family activities were inseparable. Ariès' now classic history of childhood includes heartwarming images of family "togetherness." For example, he describes a *Book of Hours* illustration in which a merchant's son helps his father with business accounts at a table in the middle of the same room in which the wife attends a younger child; bales of merchandise and domestic accouterment are intermingled. Ariès, for one, does not see a contradiction between the apprenticing of young children and the integrated nature of family life that he praises; he reports that young apprentices were treated as if they were the natural children of the families to whom they were sent (Ariès 1962; Peter Laslett 1977; Myrdal and Klein 1956).

Seen through late twentieth-century eyes, the positive dimension of these arrangements is that children routinely interacted with adults around the necessary activities of daily life, learning naturally from their elders, in their own homes or as apprentices in the homes of others. In contrast to this romantic picture of men, women, and children living and working within sight and sound of one another, the current concern about family life, reflected in the interest in flexitime, is that adults and children now go their separate ways much of the time. Everyone is in school or on the job most of the day. Thus, family members spend relatively few of their waking hours with each other. More particularly, the survival of the individual members of the family no longer depends on the sharing of tasks. The reduced need for interdependency among husbands, wives, children, grandparents, and so forth is blamed for impairing learning and eroding the functional unity of the family. This view is reflected in a National Academy of Sciences analysis of problems with contemporary American families and child rearing:

> Studies of child socialization and development in other cultures point to a distinctive feature of child rearing in America and Western Europe: segregation, not by race or social class, but by ages. Increasingly, children in America are living and growing up in relative isolation from persons older and younger than themselves. Altering this trend will require changes in American family lifestyles . . . to bring children and adults back into each other's lives.
>
> National Academy of Sciences 1976, p. 39

Few accounts of earlier times provide enough specificity about the actual division of family responsibilities—and how much different sexes and different generations actually worked together on projects, as opposed to working in separate but proximate and related domains—to allow definitive generalizations across time and place. In most societies, while women's work has included processing and preparation of food and clothing, and household care and repair, recent social histories and anthropological data suggest that only care of infants has been universally assigned to women exclusively. Otherwise, women have done practically any work that men do.

Nevertheless, it also seems that in each culture and setting the work has generally been divided by sex, so that typically, men have done one set of chores and women another. In circumstances where physical strength has made a difference, men usually have been assigned heavy tasks and women light ones. To the extent that the activities took place in different, although nearby locations (e.g., kitchen not barn, or garden not field), men and women actually may not have interacted much in their work days. It may have taken as long to walk a few miles across planted fields as to commute by car for longer distances in urban traffic. Moreover, to the degree that women were caring for young children while doing other work, the work had to be feasible with young children around; work that less easily accommodated the presence of young children was more often done by men.

Thus, to the extent that work and family lives may have been intertwined, as characterized by Ariès and others (and still are in some circumstances), the question of how much time should be spent in work, as opposed to family life, did not clearly arise in earlier eras. Presumably, flexibility between work and family time automatically exists when the settings are one and the same. Yet it may also be true that these sketches of genial generational interaction may not have been the happy "learning experiences" that they seem to some historians now.

Stage Two: When Men's Work Took Place Away from Home

For purposes of this study, the second historical stage of interest is that in which men's work increasingly took place in factories or offices physically separated from their homes. Gradually, many of the activities that both men and women had done at home were shifted to outside manufacture—clothing, gardening, canning, baking—and at the same time, many tasks inside the home eventually became mechanized—heating, food preservation, cleaning, ironing, laundering, and the like. The home ceased to be a production unit to the same degree as before (Nye 1976, p. 9; Bernard 1974; Chafe 1976). In Marxist terms, as the feudal form of the 6,000-year-old patriarchy gave way to industrial capitalism, the labor force became primarily male (Gronseth 1978).

Characterized in this way, the separation of men's work life and home life took place over a long period of time, and for different people at different times. Clearly, those economies that remained largely agrarian were still characterized by the preindustrial patterns described above. For example, the growth of industry in nineteenth-century Europe and America did not abruptly end family enterprises of the kind depicted by Ariès, Peter Laslett (1977), and Demos (1974). As late as 1900, nearly two-thirds of all American families earned their living from the land (Elizabeth Pleck 1976, pp. 8–10).

With the separation of work and home, to whatever degree it occurred in nineteenth-century America, "a new literary genre emerged," in Demos' description, "extolling the blessings of home and hearth" as a respite from the "storm" of the work world. A highly sentimentalized woman was at its center. Women were discouraged from working outside the home—except in humanitarian reform activities, usually through the churches. They were to devote themselves primarily to creating a household haven for their men, and to rearing devout and respectful children. Men acquired special authority from performing "mysterious activities" requiring the strength and endurance to maintain the entire household. Their families could not fully understand or share these labors. They worked all the daylight hours six and seven days a week, and were with the family only for short evenings (Demos 1974, pp. 432–36).

Parsons subsequently characterized this separation of work and family as an arrangement necessary and beneficial to both institutions, thus adding a twentieth-century rationale to the nineteenth-century ideal of separate "spheres" for male and female contributions to families. In Parsons' formulation one partner worked at home to provide emotional or expressive sustenance; the other worked away from home to provide economic or instrumental sustenance (Parsons 1947, 1959; Parsons and Bale 1955). In this concept of the sex role division of family labor, conflicts between work time and family time should not, by definition, arise. Despite the different realities of nineteenth- and twentieth-century female employment, this formulation of appropriate male and female behaviors and family functions has persisted, and is reflected in laws such as those determining the hours of work.

Even while stay-at-home values for women's lives were extolled—and women could not hold property and were excluded from many kinds of employment—in fact, there was a blatant double standard for women's work. By 1900, more than 40 percent of all non-white women and almost 20 percent of white American women were in the labor force. Recent immigrant and black women comprised the majority of these workers. They labored mainly in domestic service, in farm labor, and in New England textile mills. Similarly, widows and wives of disabled husbands worked for pay. The children of such women often worked with their mothers in factories and fields. These women worked because they needed the income to survive; welfare and other social services did not exist (Chafe 1976, p. 9; Smuts 1959; Bane 1976, p. 77; Gutman 1976; Rothman 1978; Scott and Tilly 1975).

Several reasons may account for the fact that the hardships of these women's roles did not stimulate social reforms, such as flexible work schedules, or more substantial changes. Very few working women were married in the nineteenth and early twentieth century (only 15 percent of those employed in 1900); and presumably even fewer had minor children (no census data on this were kept until 1940; Hoffman and Nye 1974, p. 3). Most significant is the fact that employed women had to work for economic survival, whatever the terms of the employment. The later entry of large numbers of educated middle-class women with children into the labor market created an articulate female constituency that was relatively powerful by virtue of increasing numbers, status, and resources. Perhaps equally or more important, however, this group of women was intimately tied to even more powerful men, who were willing to take up some of their causes—for example, legal efforts towards equal employment opportunities, "affirmative action," equal pay, and alternate work schedules.

Stage Three: When Women with Children Were Labor Force Participants

The third stage of work-family connections is characterized by demographic drama. In the mid–ninteenth century, life expectancy was less than fifty years. By the mid–twentieth century, female life spans averaged nearly eighty years. During the first half of the nineteenth century, women typically gave birth to eight to ten children, half of whom died in infancy. By the mid–nineteenth century, the birth rate dropped to five births per mother. By the mid–twentieth century, mothers bore an average of only two children apiece. Even in the "baby boom" of the 1950s the average number of children was only three per woman, and dropped shortly thereafter, thus continuing the general pattern of steady decline in the birth rate since 1700 (Taeuber and Sweet 1976, pp. 34–35).

With respect to our interest in the ways working conditions affect people's ability to rear children, it is important to emphasize that the decline in fertility has not been due to an increase in childlessness among women, which would tend to obviate the issue of how working mothers can care for their children. Rather, the decline simply has been in the total number of children in each family. Most women are still having children, but their age at marriage and their age at the birth of the last child have dropped dramatically. By 1973, three-fourths of married women living with a spouse had no children under eighteen years of age at home. Thus, during this third stage, women have had increasingly long stretches of adult life without the responsibilities of child rearing (Bane 1976, p. 8; Hoffman and Nye 1974; Taeuber and Sweet 1976, p. 50; Chafe 1976, p. 26; Moore and O'Connell 1978).

While these marriage and fertility trends were taking place, the proportion of adult women in the labor force rose steadily in the first half of the twentieth century. In 1900, one in every five women was in the labor force; in 1940, one in four; in 1950, one in three; and in 1978, one in two. Although there are differences among age cohorts in each period because of the development of nearly universal secondary schooling, the pattern generally remains the same (Taeuber and Sweet 1976, pp. 50–51; Grossman 1979).

Before 1940, the vast majority of employed women were young, single, and poor. Most historians writing about women's employment have assumed that World War II provided the "crucial catalyst" that increased married women's employment. Recently, however, Bennett and Elder (1979) have suggested that the Great Depression was a more important impetus. In either case, three-fourths of the women who joined the labor force in the war effort were married, and 60 percent had children of school or pre-school age. Despite the shrinkage in the number of jobs after the war, in 1960 twice as many women were employed as in 1940. In 1940, only 15 percent of married women were in the labor force; by 1950, 25 percent; by 1974, 42 percent. Finally, by 1978, 48 percent of all American wives worked at paid jobs (Chafe 1976, pp. 15–19; Bureau of Labor Statistics 1978).

The demographic reasons for the growing interest in work scheduling were accompanied by ideological shifts. Even when women's "proper place" meant domestic obligations, there were changing expectations about what constituted an ideal mother and wife. Rothman (1978) has perceived four stages in cultural attitudes and policies toward women, family, and work from the Civil War to the present. While her schematization is useful for identifying various prevailing expectations for women, in fact, each of the ideals and types persists through subsequent periods. They include:

Stage One. From 1870 to 1900 new work and educational opportunities—exemplified in the rise of women's colleges—intruded upon the precepts of stay-at-home motherhood.

Stage Two. From 1900 to 1920 the idea of "educated motherhood" emerged, focusing on the needs of the child, and reflected in Progressive era reform efforts to remove children from the workplace and to improve conditions for those who stayed. Women moved from private, philanthropic enterprises to political action.

Stage Three. From 1920 to 1950 the ideal of wife-as-companion, as opposed to wife-as-mother appeared; the ideal woman moved from the nursery to the bedroom. The career of Margaret Sanger and the birth control movement witnessed the value shift.

Stage Four. From 1950 through the 1970s the ideal of "woman as a person" developed. Women's fulfillment required not just the roles of wife and mother but expression in terms of their own accomplishments. Women's rights were emphasized over needs—rights to employment, to child care, to abortion, etc. (Rothman 1978, pp. 4–6).

Yet overriding the changing attitudes towards women's roles that Rothman emphasizes, and as the Rapoports and their coauthors (1977) and others have pointed out, are Parsons' views about appropriate roles for men and women, which have persisted in the twentieth century as authoritative strictures about appropriate child rearing and parenting. These conceptions of "normal, mature" men as economic providers, and of "normal, mature" women as housewives and mothers have been articulated in various fields: in psychiatry by Bowlby (1951, 1971, 1972, 1973) and Winnicott (1963/74); in medicine by Spock (1946); in sociology by Anshen (1959); and in law, education, and social work. The division of labor by sex was declared biologically natural and universal, therefore "functional." Not until the 1970s was this model of family life seriously questioned, leaving the decade of the 1980s with the task of establishing new models (Rapoport et al. 1977, p. 87).

Despite the reality of dramatically augmented female employment after 1940, persistent nineteenth-century attitudes about women's "proper place" were demonstrated in the paucity of arrangements made to help women handle their new roles. For example, although the government actively recruited mothers of young children to work in the war industries, the mothers themselves were still expected to bear primary responsibility for their homes and children. Day care and other community services were provided only when the women could not be induced to work otherwise—and the number of child care centers never adequately met the need of working mothers (Chafe 1976, p. 19).

Wilensky (1961, p. 50) has pointed out two striking changes in the nature of post-industrial working time, both of which affect work and family connections: first, the sequence and timing of tasks is more disciplined and ordered; and second, increased fragmentation of tasks has decreased the flexibility of daily and weekly schedules.

Significantly, however, the women entering the labor force in increasing numbers after 1940 worked in arenas that circumvented these trends. Their jobs permitted both more personal autonomy in performing tasks and more scheduling flexibility. Apart from the increased number of women in clerical jobs, and those in industrial jobs during the war, the majority of women who stayed employed or became employed after 1940 were women from the middle class who entered nursing or teaching (Chafe 1976, pp. 13, 17). In addition to the sex-segregated traditions that assigned most of such jobs to women, the fact that American schools usually dismiss in mid-afternoon has allowed mothers of school-age children to teach and still be at home when their own children finish the school day. Nursing has offered another kind of flexibility—namely, shift work—that lets women be at home with the children.

Choosing work with schedules that are easily adjusted to family responsibilities has generally forced women to accept modest career goals. Despite their growing numbers in the ranks of the employed, women have not proportionately increased their presence in supervisory and professional positions. In addition to well-documented discrimination on the basis of sex alone, another important reason why women are not distributed proportionately throughout the work hierarchies doubtless has to do with the hours they are able to work, given the concurrent requirement that they also care for the home and children. In professional work, the ability to spend longer hours on the job, and to keep irregular hours when necessary (as school or health care administrators or physicians, for example) is usually correlated with promotion to higher levels of responsibility, prestige, and pay. Women's responsibilities at home have systematically precluded this kind of involvement with work. Similarly, taking "time out" from working to bear and rear children has set women back in the fields in which continuous involvement in work is necessary for both substantive and organizational advancement.

It is the increased employment of women with children that has given major impetus to the current focus on work-family connections. In 1978, over half the mothers with school-age children were in the labor force. In addition, 49 percent of women with preschool children ages three to six years and 34 percent of those with children under three were employed. Two-thirds of these mothers worked full-time. Among the mothers with children under three, 33 percent of those with husbands present were employed and 49 percent of those ever married were employed (Grossman, 1979). Many observers think that many more full-time housewives (40 percent) would seek work if attractive opportunities existed. Growing numbers of women are enrolled either in further education or in training programs—with the expectation of future employment (Sexton 1977, p. 19; Sawhill 1977; Bronfenbrenner 1977a; Presser 1978; Berkove 1979; Hooper 1979).

Even with the long range decline in birth rate in the United States since the beginning of the nineteenth century, most American men and women still marry and have children. Current projections are that 90 to 95 percent of America's women in their twenties are expected to marry at some point (Glick 1977; Moore and Rogers 1979).

But two additional emerging child-rearing trends are related to the current attention to work-family issues. The average number of children per family dropped to two in 1977. Furthermore, many observers think that more college-educated women are waiting until their late twenties and thirties to have children. Although there are no national data on this trend yet, the average age of American women at the birth of the first child has risen slightly in the last decade (from 21.8 years to 22.6 years; Presser 1978).

The divorce rate has doubled in the last ten years, and since the remarriage rate for women is lower than the divorce rate for families with children, the result has been an increase in single-parent households (Waldman, Grossman, Hayghe, and Johnson 1979). One in every six children under eighteen years of age now lives with only one parent—usually a mother. Three-fifths of these single mothers hold jobs, usually full-time.

Although less widespread than the trends described above, the escalating number of women aspiring to, or already in, high-level, high-pressure, and time-consuming careers also has implications for American family life. Ten years ago, 5 to 10 percent of law school and medical school students were female; today 25 percent are women. The number of women lawyers and physicians almost doubled between 1960 and 1970. The number of female college teachers tripled in this period (Bureau of the Census 1976; Farr 1978).

Concurrent with the rising average age of the population, three additional work-family issues are apparent:

- Large numbers of older people wish to remain in the labor force on a flexible or reduced time basis.
- Increasing numbers of employed adults need time to spend with or help care for older members of their families.
- Students of all ages seek part-time or flexitime work to en-

able them to care for their families, to attend school, to get job experience, or to help pay rising educational costs—or often a combination of all three.

In a more theoretical vein, but equally relevant to work-family issues, is the current scholarly reexamination of our culture's assumptions about the critical and continuing nature of family influence on the developing individual. Some observers object to the phenomenon of extra-familial institutions in the society—schools, churches, social service organizations—replacing or supplementing families in caring for many family needs (e.g., Lasch 1977). Others question the importance of parental influence in comparison with other social and economic factors, as well as whether early influences are as indelible as currently believed (Kagan, Kearsley, and Zelazo 1978). The conflict of perspectives suggests the importance of comparing different work schedules with respect to how they affect parental time with children.

Conclusion

The historical trends outlined above have dramatized the connections between work and family activities, and have raised concerns about the extent to which work systems allow people to maintain effective participation in both work and a family. The underlying assumption is that the developments described, whether taken separately or together, have complicated the challenges employed people face in balancing their work and family involvements in satisfactory ways. We need, then, to ask:

- What happens to women when they have jobs and families?
- What happens to children when their mothers, as well as their fathers, work away from home during most of the children's waking hours?
- What happens to men as a result of the changes for women and children?

- How can a society successfully support its employed adults in their need to care for both their young and their old? (Increasingly, too, with most Americans living to seventy and eighty years of age, employed parents must provide various kinds of support and care for their own parents.) What structures and attitudes should change?

The next chapter discusses the answers to these questions that have emerged from national discussions of alternate work schedules in the last half-decade. Our analysis of the answers suggests that despite the prevailing optimism and apparent agreement about the benefits of flexible work schedules for families, expectations as to how they will help families, in fact, conceal differing cultural values about how work and families should be related in American life. As will be apparent in the discussion, these differing views are legacies of the changes in, and expectations for, families in America's early years.

3

Consensus and Controversy about Flexible Schedules

IN the last five years, the major discussions in American public life of the expected benefits of flexitime for families have taken place in congressional hearings on the bills to expand flexitime and permanent part-time opportunities in the federal service.[1] The expectations of the sponsors and supporters who favor flexitime for family reasons are that greater flexibility in various kinds of work schedules will make it easier for people to manage both job and family responsibilities. Given this proposition, flexitime and part-time can be seen as a continuum, with permanent part-time work representing an additional extension of the principle of providing more choice for individuals in order to allow them to accommodate multiple demands and degrees of interest in their work and family. While the hearings are not the only arena in which the connections between the timing of work and family functioning have been addressed, the testimony reflects the ways in which the issues have been discussed in various settings.[2]

[1]The flexitime and part-time bills were introduced and debated in tandem because they serve essentially similar goals. Virtually identical arguments—from the standpoint of family issues—were advanced for both. See U.S. Congress, June and Nov. 1977; Sept. and Oct. 1975; April 1976; May, June, July, and Oct. 1977; and Feb. 1978.

[2]See U.S. Congress, Hearings 1975, 1977, 1978, for copies of the bills and

Among the most important themes to emerge from our examination of the hearings is that a variety of values and expectations for family life coexist beneath the enthusiasm for flexitime. This fact has important implications for both whether and how work schedules ultimately will help people balance work and family life, as well as important implications for the process of attempting family impact analyses of public policies. (See further discussion of the issue of differing values about families in the concluding chapter.)

The groups praising the family benefits of flexible work schedules together represent an array of current ideas, worries, and hopes about children. Arguments in favor of flexibility in work schedules oscillate between concern for the well-being of family members as individuals who are required to play more than family roles, and concern for the members' role in the family vis-à-vis each other. Interest in the second instance focuses on the effect a person's performance in the role will have on other members of the family, and therefore on the unit as a whole. Finally, interest in the relationship between families and work schedules invariably is tied to a general assumption that there are connections between the well-being of families and the well-being of the society as a whole.

---

testimony on flexible and part-time legislation. See also Whittaker (1978) for a legislative review. In addition to the congressional hearings, the Government Accounting Office, a congressional office that independently investigates executive branch activities, prepared a report on flexible and part-time programs in the federal service (see U.S. Comptroller General, 1977). Most other published reports and articles on flexitime in the U.S. have appeared in professional journals, e.g., L. Smith 1977; Martin 1974; Legge 1974; Langholz 1972; Elbing, Gadon, and Gordon 1974; Evans 1973; Glickman and Brown 1973; Swart 1978; Zagoria 1974. Occasional feature stories have appeared in the general press, e.g., Nathan 1977; Stein, Cohen, and Gadon 1976; Fellows 1971; *Business Week*, "Europe Likes Flexi-time Work," 1972; *Business Week*, "Flexible Hours," 1973. Various interest groups have also prepared papers on the advantages of flexitime for constituent groups, e.g., Flynn 1977; National Council for Alternate Work Patterns 1977.

The first and broadest of the several sets of cultural values and interests that lie behind the respective arguments is concerned with the quality of work in general and with efforts to "humanize" the workplace. As heirs to the various reform traditions aimed at worker well-being, the current advocates consider alternate work schedules as one of many possible ways in which workers may gain more control over their lives. They propose that workers have a say in how their work life takes place. Giving workers choice, or flexibility, in when and how much to work is said to constitute an important step towards an increase in their feelings of autonomy and control. The assumption is that with wider choices, people can create more satisfying, diverse, complete, and rounded lives. The corollary to this is that the multiple capacities of people are likely to flourish to the degree they feel they have control over, and therefore many options in, their lives.

Many independent organizations expressed these views at the three congressional hearings that were held from 1975 to 1978. In addition, the various sponsors of reform legislation during the five years it was debated in the Congress invoked the same perceptions of human nature to support greater employee choice in work schedules. For example, in support of his flexitime bill, Representative Solarz explained that,

> One of the real problems we have in this increasingly mechanized age in which we live is a growing feeling of alienation on the part of many people. They sense that they lack the ability to control their own destinies. I think that one of the great advantages of flexitime is that it gives an individual an opportunity to play a decisive role with respect to one of the most critical components of everyday existence, which is the hours of work. And to the extent we give individuals a greater sense of mastery over the conditions of their employment, over the conditions of their existence, as it were, I think we deal with one of the most significant psy-

> chological problems which we face in our increasingly industrialized civilization.
>
> Solarz 1977, pp. 7–8

And Ersa Poston, a commissioner of the Civil Service Commission speaking for the Commission in the most recent Senate hearings in favor of the flexitime legislation, echoed Solarz' view: "We regard the opportunity to have some control over one's own work time as an important step forward in improving the quality of work life" (Poston 1978, p. 32).

Apart from such broad, philosophical statements about the value of autonomy and control, the discussions of alternate work schedules in the congressional hearings divide into three roughly equal parts. Approximately one-third of the questions and answers deal with implementation issues (e.g., supervision costs, the mechanics of time keeping, and the pros and cons of mandatory programs and quotas for part-time slots). Another third of the testimony concentrates on the advantages of alternate schedules for employers and wider social purposes (e.g., to reduce the need for more energy, more highways, more recreation facilities, etc.) A final third of the testimony addresses the advantages of flexitime and part-time schedules for categorical groups—older workers, students, and the handicapped, and parents, particularly women—in relation to their roles as family members.

Family-related Arguments for Flexible Schedules

Although the reasons for favoring flexible work scheduling are often listed without reference to family issues, in fact, behind many of the specific problems employers hope to alleviate with flexitime (e.g., tardiness, absenteeism, low morale) are conflicts between time for work and time for family members. The conflicts identified are those most often felt by women. For example, various congressional witnesses recalled that the flexible schedule idea was originally thought up by a German female economist who was looking for ways to increase female labor force participation during a labor shortage in Germany in the early 1960s, and that her concept was explicitly designed to provide a way for women to work when they also had family responsibilities (see the discussion of Christel Kamerer's idea in Chapter 4).

According to testimony in the hearings and to employer evaluations, subsequent experiences with flexitime in the United States and Europe confirm her expectations. Tardiness and absenteeism, which flexitime appears to reduce dramatically, apparently were due largely to the necessity for women with families to take minutes—or days—off work for their children's needs. When employed mothers are given more leeway to allocate their time for work or family members as they need to, tardiness and absenteeism diminish. Morale at work tends to rise when worries about meeting both children's and employer's needs and expectations are reduced.[3]

The need to help women, particularly mothers, is the most frequent family-related argument in favor of flexible work schedules. The same reason underlies support for flexible schedules on the part of women's lobbies, such as Women's Equity Action League, National Organization for Women, National Women's Political Caucus, Women's Lobby, and various state commissions on the status of women. Female members and ex-members of Congress—including Yvonne Burke, Bella Abzug, Gladys Spellman, Patricia Schroeder—also support the legislation for these reasons—as do "expert" witnesses from various fields, e.g., labor economists, mental health professionals, and representatives from business and government agencies that utilize flexitime (Ralph E. Smith 1977b; Owen 1975; Haldi 1977; Eyde 1975, p. 151; Social Security Administration 1974; Shuck 1977, p. 154; Finegan 1977, pp. 155–57).

[3]For example, Shuck 1977, p. 154; Marsh 1977, p. 102; Martin and Hartley 1975; Cowley 1976; National Geological Survey 1976; Social Security Administration 1974.

Who Needs Help Balancing Work and Family?

The following vignette, describing an employed mother coming home from work and attending to her family and household, suggests the nature of work-family "balance" problems. The story illustrates both the substance and intertwinedness of child rearing and home chores, and the stress related to managing these responsibilities for employed people in metropolitan areas, especially parents, and more especially mothers.

In the example, a thirty-four-year-old woman has two children, three and six years old; she works as a secretary in a downtown Washington, D.C., federal department and commutes from the suburbs, as most federal workers do. Her husband is a program analyst in another federal agency in a different part of the city. This family structure—that is, two employed parents with young children—is one of a range of contemporary family types; it represents one of the most taxing and "continuous process" life cycle stages.

> A few minutes after 5 P.M. Darlene tries to walk quickly from her office building to the subway stop two blocks away, but the sidewalks are crowded with hundreds of other mid-town Washington workers also eager to get home. She has to wend her way slowly, trying not to bump into people coming the opposite direction. She stands in line to go through the subway turnstile. Her train arrives within a few minutes; but the cars are crowded and she has to wait another five minutes until the next subway comes.
>
> She rides fifteen minutes to the stop where she catches a bus. She managed to get a seat on the subway but has to stand for the additional twenty- to twenty-five-minute bus ride. Her legs begin to feel the accumulated fatigue of the day. She begins to worry about how she will get everything done that evening.
>
> The bus brings her within a ten-minute walk of her

babysitter's house, but tonight she won't be walking. Because her husband rode the bus to his job this morning, she was able to leave the family car in the parking lot close to the bus stop. Now she can drive to pick up the children, and then stop at the grocery store to get milk, bread, and tuna fish on the way home. She would rather not take the children shopping with her at this time of day because the store is crowded and the children are tired and hungry. But the family needs the groceries for dinner, breakfast, and tomorrow's lunches.

Inside the house, Darlene takes off Betsy's winter jacket and reminds John to hang up his coat. It is almost 6:30. To ease the characteristically urgent late-afternoon hunger of a preschooler while she prepares dinner, Darlene hurries to the kitchen to give her daughter a cracker, carrot, or whatever she can find quickly. As she puts away the food and takes out items for supper, Darlene hears about first grade reading and who did what at the babysitter's. John turns on the television and watches intently.

Tonight Darlene's husband, Sid, takes a class and will be home at 8:00 instead of 6:30. Betsy tugs a chair to the counter to help stir. Darlene asks John to get napkins from the drawer. She and the children eat together. As Betsy reaches for her spoon, she knocks over the milk cup. Thus, before Darlene can take a first bite, she must jump up from her chair and rush to the sink for a sponge to wipe the spilled milk off the floor, the table leg, and the chair. She helps Betsy out of her wet shirt and goes to the bedroom for a dry one. Sitting again at the table she reminds John to try the vegetables. Betsy begins to slide off her chair. Darlene tells her not to get off her chair again or she won't get dessert. It is 7:15.

After the meal, Darlene makes several trips from the table to the sink with dirty dishes, puts away the food, and starts water running in the tub for the children's bath. While the children are bathing, she collects a load of clothes from the hamper and the children's rooms, puts it in the washing machine, and folds the load which was still in the dryer from yesterday. A phone call reminds her she has promised to bake something for John's class party at school tomorrow. It is after 7:30.

John calls from the bathroom that they are out of toothpaste. And the last button has come off his pajamas. After the children's bath, Darlene starts to read them a story from a library book. A phone call interrupts her, asking if she can sit for the babysitting pool this weekend. As she finishes the story book, both children beg her to read another. She hugs them, saying it is too late and that they must get to sleep.

Her husband arrives home just then. It is 8:00. The children jump up to greet him, both talking excitedly at once, hugging him. They persuade him to read another story. While he gets ready—takes his coat off, gets something to drink and a snack—they run through the living room and kitchen, chasing, playing.

Darlene finishes the dishes and picks up the bedrooms. Sid reads to the children for ten minutes. It is 8:30. Betsy protests going to bed. She says she needs a drink of water. Sid gets her a drink, carries her to her room, kisses her goodnight. Finally, she is settled in her crib, lights out. Darlene goes to kiss John goodnight and talk for a few minutes to him. Betsy's door opens. She is out of bed and wants another drink of water. Darlene and Sid make four more trips in and out of Betsy's room with sterner, crosser admonishments each time. Finally, Betsy sleeps.

The new load of laundry has to be dried. And folded. And put away. The school cake must be baked. The bathroom sink and tub need scrubbing. Cracker crumbs litter the hall rug. Vacuuming hasn't been done for five days. Both Darlene and Sid are tired. Their office days had been pressured by deadlines. Darlene feels she was not as patient and loving with the children as she likes to be. The mess in the house is getting her down. She had hoped to talk to her friends Mary and Ruth tonight. And she wanted to write to her parents. She warms up dinner for Sid, mixes the cake, and hears about his day.

Sid turns on the television and watches while he writes checks for a couple of bills. Darlene does a few more chores: makes lunches, starts dinner for tomorrow, irons a blouse, puts new laces in Betsy's shoes, and writes a permission slip for John's soccer game.

At 11:00 the parents go to bed. At 3:30 A.M. Betsy wakes up crying—a bad dream. Darlene and Sid each get up once to hold and comfort her. Betsy finally falls back to sleep at 4:10.

At 6:30 A.M. the alarm goes off to start the next day. Darlene gets up to ice the cake. She gets the children dressed and fed before she leaves to catch her bus at 7:30. Sid will take Betsy to the babysitter on his way to work. John walks to school with a neighbor.

For our research—on a version of flexitime in a sample federal agency—the question is whether or not giving workers, especially employed parents like those characterized in the vignette above, some choice about when to work their eight hours each day helps them to deal with family needs—child rearing, chores, and family management in general. Several aspects of the issue seem important.

First, for parents, child rearing and home chores—particularly during the late afternoon, supper, and the bedtime hours—

are inextricably intertwined, especially for families with young children. For families with teenagers, the chores and rearing may be interrelated in slightly different ways. The teens may "hang around" the kitchen and talk and help with dinner in the late afternoon, whereas they may have homework or other activities at a later hour.

Second, mothers (employed or not)—not fathers—are the parents primarily responsible for the multitude of interrelated daily activities required to sustain family life (see the discussion of time budget data in Chapter 4 and Appendix D).

Third, there is an interrelationship between the amount of time available to do the chores and to be with the children and the particular segments of time available. What if the mother in the vignette arrived home at 4:30 instead of 6:30—would she spend more time with the children? Could she feel more relaxed and patient about having the children participate in shopping and dinner preparations? Or about stopping some of her "doing" to look at something they wished her to see, or to help them with a project? Does exactly when she has that "extra" time matter? Does having the time matter? What would the differences in her availability and mood mean in the children's lives?

Fourth, if a father has the option of flexitime, would he take advantage of the chance to get home earlier and share more of the routine child rearing and home chores with his wife? Would he engage in additional activities with the family—helping with homework or projects, playing or talking with other family members? Pleck (1978a) notes that despite the central importance of the presence of children in increasing the role demands in two-earner couples, it has been one of the most difficult components of family work to conceptualize for research purposes.

Does flexitime also have advantages for people without children, with different daily family routines from those described in the vignette? Would it allow them to take time over lunch to go to the post office, or shopping? Or to leave late in the morning, or get home early for a repairman, a delivery, or recreation?

Or to shop or do an errand for another family member during the working day—any of the small but ever-necessary family-related activities?

Concerns of Employed Women

The major concerns raised in the hearings for employed women in general and mothers of young children in particular are of two kinds: first, work overload and, second, curtailed employment opportunities.

Work Overload

Employed women carry a double work load, more so than most men, because they take primary responsibility for managing their daily family activities, as well as for working and commuting. The total number of hours spent on both job and family exceeds most men's hours in these two areas. For example, a career consultant to women reentering the labor market testified that,

> Women seeking employment also share a commonality of pressures which rarely affect men. First, she is expected to maintain her household regardless of the constraints of outside work. While it is true that some men voluntarily assumed some household chores, social expectation still emphasizes the women's role. Employers still expect mothers, not fathers, to take time off when children are ill. Executives still expect their wives to entertain. Psychologists and other professionals continue to stress mothers' rather than fathers' influence on children. Thus, working women are usually required to assume responsibility for husband, children and home. Even when they do not, employers assume that they will.
>
> Melia 1975, p. 74

Echoing Melia, representatives of the Women's Equity Action League and Women's Lobby explained that,

> On a conventional full-time work week schedule, the working mother, married or single, has to depend upon healthy children and schools being in operation. She has to hope the children don't need to go to a dentist, a doctor, or any kind of therapist. Shopping has to be done at night, or . . . on weekends. . . . And if the children are to sing in the school Christmas program, it had better be held in the evening or the parent cannot attend.
>
> Hendrickson 1975, p. 73

Within this general understanding of the nature of employed women's dual roles, the expressions of concern can further be divided into two categories:

1. Those who assume that it is appropriate for women to carry most of the child rearing and domestic responsibilities mentioned above, even when they are employed, but who also feel women's time conflicts should be eased through work schedule flexibility.
2. Those who think both sexes should have increased work schedule flexibility for the sake of family needs, and in order that inequities between males and females in family and work can be evened out.

The first view is seldom expressed explicitly, yet it is usually implicit in statements supporting women's labor force participation, as in the following examples:

> The flexitime systems make increased female labor force participation more feasible, partly because it enables the working mother to provide better care for her children.
>
> Owen 1975, p. 58

> Under a fixed work schedule, many women are virtually excluded from working because of domestic commitments, particularly the care of young children. The flexibility afforded by a flexitime system will allow women a greater opportunity to accept outside employment, a condition which should contribute substantially to the achievement of society's goal of equal rights for women.
>
> Kuhne and Blair 1978, p. 43

> The change [flexitime] pays off for almost everyone, but most for working mothers.
>
> Nathan 1977, p. 37

In these views, flexitime schedules are desirable simply to ease the complications of what mothers have to do—not because they may allow fathers to share domestic demands with employed mothers.

With respect to the second interpretation of how work schedule flexibility might ease women's dual roles, a perceptible shift in emphasis occurred in the arguments during the five years the bills were debated. In the early testimony, the needs of mothers and children were cited almost exclusively. Men were mentioned only in terms of how they might find advantages for themselves in spending more time with their families, as in the following examples:

> While women would gain substantially from enactment of this legislation . . . there are men who wish to assume a greater role in child rearing and family life.
>
> Burke 1975, p. 17

> I think for working fathers it is important because . . . they may want to phase down some of their work so they can be on the family scene more. It would give fathers also more options than they normally have.
>
> Schroeder 1975, p. 21

In later testimony, expectations about men's roles became more defined. Men were mentioned increasingly in terms of their obligations to share in home responsibilities, as illustrated in the following exchange between the congresswoman from Colorado and a member of the Colorado Governor's Commission on the Status of Women.

> *Ms. Lamm:* I'm really glad to hear all the other testimony today because before today I've heard mainly that this is a woman's issue. As I get more and more into it, I see and feel very strongly it should not be only a woman's issue. I certainly would not want it to be used to have women working harder on their jobs and at home.
>
> *Congresswoman Schroeder:* The homemaker adds thirteen hours a week to her working when she goes to work full-time. The average male when his wife goes to full-time cuts back his hours by about two a week. Something is wrong.
>
> *Ms. Lamm:* That's right, and I would like to see this as a family-concerned issue, and I would really look to see if it is used more to help men and women share the responsibilities at home.
>
> Lamm 1977, p. 107

Psychologist Bronfenbrenner of Cornell made a similar point:

> [We should] make the role of parent a more acceptable role in our society than we now do . . . for a man as well as for a woman. . . . I'm very glad this legislation does not specify sex as one of its criteria.
>
> Bronfenbrenner 1977a, p. 30

Gradually, discussion of work-family conflicts were framed in terms of equity for mothers and fathers. Witnesses increasingly argued that the laws of the society should recognize and support symmetrical roles in work and family for men and women, and for the sake of the children. Flexitime and part-time opportunities were supported as catalysts to these work-family realignments:

> [With flexitime] two parent families could change schedules and allow fathers to spend more time with their children and share the homemaking load.
>
> Lynne Revo Cohen 1978, pp. 82–83

> Since fathers would have flexitime opportunities as well, the care of children could be shared to the greater benefit of the whole family.
>
> Gladstone 1977, p. 33

> Flexible hours allow for parents, men and women, to share responsibilities as well as the joys of parenting. More time for family is the most significant reward of these bills.
>
> Renteria 1977, p. 131

With reference to Swedish laws allowing parents to work part-time (six hours a day while their children are small), Carol Greenwald, former commissioner of banking for Massachusetts, described the dilemma of combining parenting and working. She argued that in the United States,

> What has not been as clearly worked out is how the children are to be cared for in this new family setting of mothers and fathers participating in the paid labor force. Society has an important stake in preserving the family unit and ensuring that children are well cared for—physically and emotionally. While surrogates can and should play increasingly important roles through increased availability of child care centers and after school programs, parents must continue to devote substantial time and energy to parenting for both the child's and parent's emotional well-being. Society's in-

> terests coincide with those of families in allowing parents greater flexibility in their work modes so that they can also parent. The widespread adoption of part-time work options during the child-raising years for both parents would offer an excellent means of providing for the emotional and other child care needs of children without sacrificing the careers of women or the efficiency needs of the economy.
>
> Greenwald 1978, pp. 45–46

In response to a question from Senator Thomas Eagleton about what should happen when the supervisor for whom a person works says, "I want my stenographer here eight hours a day, five days a week," Greenwald replied that public policy should take the position that "we have a great deal of concern about the care of our children, and our concern for their well-being overrides your concern for the convenience of doing dictation between nine and ten in the morning" (Greenwald 1978, p. 39).

In the House hearings, Bronfenbrenner posed the work-family value questions for American adults in even broader terms: "We are worried about bureaucracy and civil service regulations that regard a person who is willing to work full-time at regular hours as more worthy, more important to the society than a parent who seeks a flexible or part-time schedule in order to be able to care for his or her children" (Bronfenbrenner 1977a, p. 31).

Curtailed Employment Opportunities

The second major concern expressed for employed women in the discussions in flexible work schedules is that their family responsibilities often interfere with their work lives. Frequently women, especially those with young children, choose to work part-time or take time out of the labor force for months or years. They pay an "opportunity cost" for this time off in the sense that they are at a disadvantage when competing for work opportunities (e.g., in career advancement, in accruing fringe benefits, and in securing new jobs) with men and women who do not take comparable time off from their jobs. Flexible schedules are seen as a way to help women around this problem: "The increased degree of personal freedom characteristic of flexible schedules could be particularly important to working women with child care responsibilities. Individuals who have not been able to enter the labor market because family demands made conformity to traditional work schedules impossible may now find this obstacle removed" (Spain 1975, p. 28).

In her testimony, Congresswoman Schroeder stated the hope that flexible work schedules would allow women to sustain their careers while raising children:

> I think very often it has been because of the lack of flexible hours and part-time ability that women have been forced to drop out rather than stay in [the work force]. It becomes a crunch between the husband's career and the children versus keeping your finger in the dike and the woman is the one who yields because of society; so as a consequence when you go back in you have lost the training years, you have lost seniority, you start back again, and you just never quite are able to move up on the career ladder in the same manner.
>
> Schroeder 1975, p. 22

Gladys Hendrickson, representing the major national women's lobbies, pointed out additional costs to women of temporary absence from the labor force:

> With a withdrawal from the labor force, women lose all the fringe benefits that are so important to workers: credit for those withdrawal years in Social Security and pension plans and insurance programs. They also lose touch with job contacts and new trends in the job market. The loss of those fringe benefits during the with-

> drawal years, and the poorer jobs found at reentry are reflected in the particularly low incomes of many older women. . . . Currently women are recognizing that with increasing uncertainties about the future it is important to have marketable skills and, as far as possible, to be economically self-sufficient. So instead of facing the problems of withdrawal from the labor market and reentry, they are trying to remain in the labor force during the child-bearing years.
>
> Hendrickson 1975, p. 71

The connection between the above issues of women's well-being—in terms of work overload or curtailed employment opportunities—and the well-being of their children is sometimes made and sometimes not. But in either case, discussions of flexible work schedules in relation to family issues eventually address the presumed advantages in terms of children. The explicit, or implicit, assumption is that flexible working hours give parents—usually meaning mothers—more time and less stressful time to spend with their children; more and less stressful parental time with children is implicitly a "good" thing:

> What does it mean for children . . . that more and more mothers, especially mothers of pre-schoolers and infants, are going to work, the majority of them full-time? Paradoxically, the most telling answer to the foregoing question is yet another question: Who cares for America's children?
>
> Bronfenbrenner 1977a, p. 27

> We are facing the major revolution and change in American society, far outdistancing anything the energy crisis has to say to us, and that is the changing structures of our families. . . . We have for the first time, more women working full-time than was true at the peak of World War II. We have more families where

> both are breadwinners, and we haven't stopped to say what does that mean in terms of the very basic questions of who is going to care for our children?
>
> Greenwald 1978, p. 38

> Flexible working time provides [parents] a chance to spend more time with families. This is especially beneficial with school-age children. They will be able to schedule their time so they are home when the children return from school.
>
> [With flexible and compressed work schedules,] working parents would find it easier to meet both family and work obligations. Parents working under flexitime could arrange their schedules around the opening and closing times of child care centers. . . . Children would have more parental care and attention than is now possible. For working parents, the tension caused by trying to meet conflicting commitments would be eased.
>
> Sullivan 1978, p. 278

Finally, most witnesses in the congressional hearings also mentioned other categories of family members who would be helped by flexibility in work scheduling (particularly part-time positions), namely, older people, students, and the handicapped.

Thus, several kinds of expectations for flexitime in relation to families can be identified in the congressional discussions, and from general articles. Foremost, the family advantages are seen in relation to employed women and the dual load they normally carry when they also have children and household responsibilities. Alleviation of the stress women feel in the double roles is seen as the major family benefit of greater work schedule flexibility, along with enhancing their opportunities to stay abreast of men in the labor force.

Second, it is expected that children of employed women will be better cared for when their mothers are under less stress from time pressures at work. Finally, some observers hope that greater flexibility will encourage men to increase their time with children and domestic responsibilities both for their own benefit, and to create more equity with women in work and family. In general, those interested in work flexibility for the sake of family well-being encourage the change as a step towards allowing more parents of both sexes to have both fulfilling work *and* family lives—for themselves and for their children.

4

Conceptualizing the Study

BECAUSE the problem under investigation in the present family impact analysis encompasses two topics that traditionally have been considered separately in research and public policy—namely, work and families—in planning the study we took into account the theoretical and empirical literature in both areas. The relevant literature spans a number of disciplines and an enormous quantity of material. Studies, reviews, and discussions fall roughly into seven categories:

1. Economic analyses of labor market issues (e.g., part-time work, female labor force participation, and the market value of housework).
2. Sociological studies of roles (e.g., in work, leisure, marriage, and in family satisfaction, power, and the division of labor).
3. Psychological and medical research on child and adult development (e.g., mental and physical health, and vocational development).
4. Organization and labor specialists' examinations of unionization, worker satisfaction, alienation, hours, rewards, stress, and quality of work life.
5. Political science and public policy studies of social services (e.g., welfare, and health care delivery).
6. Ethnographic studies of communities and whole families.
7. Historical studies of families.

(For a review of the literature in some of these areas, see Appendix A.)

In her review of the literature addressing the interconnections between family and work, Kanter (1977, pp. 7–8) argues that in spite of widespread agreement that the family and the economy are linked in broad ways—for example, that family circumstances are closely tied to the financial earnings of workers—the specific intersections between work and the family have been largely ignored in policies and treated in only a handful of studies. In her assessment, the transactions between occupations, on the one hand, and families, on the other—both as connected organizers of experience and as systems of social relations—have been virtually overlooked. Because only a few studies consider the experience of people in their family and work situations by looking at them in both contexts, she concludes that a myth of separate worlds has prevailed in both scholarly and public policy approaches to the topic.

Despite the scholarly and policy inattention to work-family interconnections, Kanter identifies five dimensions of work experience that bear on family relations:

1. Absorptiveness of the work. To what extent is the occupation highly salient, requiring both commitment and time of the employed individual and other family members?
2. Rewards. What are the material and symbolic rewards—money and prestige—of the job which determine economic and psychological security and power both within and outside the family?
3. Cultural dimensions. What are the values of the work environment and how are they taught? How do they generate a characteristic outlook on the world and an orientation to self and others?
4. Emotional climate. How does the nature of the work, and a person's placement in the organization, contribute to a worker's feelings which are brought home and affect the tenor and dynamics of family life?

5. Time requirements. How does the total time demanded by the occupation and the timing of the work affect family events, routines, and interaction?

This last dimension, time, is the realm of concern for this study. Various scholars have pointed out that time has not heretofore been an important concept in family research (e.g., Hansen and Johnson 1979, p. 589). A review of the literature related to work scheduling suggests that few existing studies have isolated work schedules as an independent variable in order to examine its effects on families.

Yet, as Kanter (1977) suggests, the sheer number of hours spent at work, as well as which part of the day those hours encompass, can influence a large number of family processes. For example, a worker's fatigue or unavailability to take responsibility for or participate in family activities may cause problems. Or work demands that extend beyond formally designated hours (e.g., overtime) may intrude on time the family expects to claim, thus causing stress for the worker and other family members.

Similarly, non-work hours for an employed family member may not coincide with the leisure hours of other family members; for example, when children are out of school but not yet asleep. Work-related travel may also disrupt daily routines, making it difficult for families to establish stable patterns of interaction. Reflecting these kinds of conflicts, the 1977 Michigan Quality of Employment Survey reported that a high percentage of respondents complained about lacking control over their work schedules (Staines and Quinn 1979). Despite the importance of these issues, little research has explored their implications for family life. In her review of research on child care in the family, Clarke-Stewart (1977) concludes that although a great deal is known about the qualities necessary for care-giving, little attention has been paid to how social institutions may be structured to support or enhance family care-giving.

Work Time

The question of how much time Americans actually work is a subject of some controversy in the literature. In general, workers in higher occupational groups work more hours than those in low prestige occupations. But within the professional groups there are variations. For example, various studies have found that lawyers and professors work longer hours than dentists and engineers (Wilensky 1961; Gertsl 1961). Also, blue-collar workers who have regular forty-hour-a-week jobs also may have second jobs for more income, and thus work longer hours (Kanter 1977, p. 32).

Taking the concept of work outside the labor market context, and writing implicitly mainly about men, De Grazia argued in 1962 (pp. 148, 167) that Americans do not work less when they have the choice; if they cut down on work in one of its guises, namely on the job, it appears "with other faces," for example, in longer commutes, in shopping, repairs, or housework. Much of what is called "free time" is another kind of work that leads to a "worthy purpose." More recently, however, it has become widely recognized that employed mothers have almost no "free time"; when their "family work" hours are added to their hours at paid jobs, their work days exceed those of both employed men and housewives (Robinson, Converse, and Szalai 1972). (See further consideration of this point in discussion of dependent variables and in Appendix D.) Recently, the "new home economics" has reexamined the traditional economic definitions of work by attempting to take into account the market value of "family work" (e.g., Sawhill 1977; Ross and Sawhill 1977).

To see the contemporary work-time issue in perspective, it is also necessary to understand the several ways in which legally mandated work hours have been steadily reduced during the twentieth century: first, by shortening of the work day from 1900 to 1940; second, by the increased use of paid holidays and paid vacations; and third, by the reduction of lifetime hours of work through delayed labor force entry (by prolonged schooling) and early retirement. Considered historically, the current definitions of full-time and part-time work lose substantive meaning and reflect instead simply the expectations of the historical moment. For example, when ten-hour days were the norm, eight-hour days would have been considered part-time.

But contrary to popular opinion, the average number of hours contributed to the labor market by American husband and wife families has increased, not decreased, since the end of World War II (e.g., Owen 1978a, p. 31). A "technology assessment" of alternative work schedules (Haldi 1977) determined that the average amount of time married couples spend working and commuting per week in 1977 was ten hours greater than it was in 1940. The combination of increased female employment and the maintenance of the forty-hour work week since the end of World War II (with no reduction in average weekly hours) has contributed to this phenomenon. Most significantly, the increase in total family hours of work is made up almost totally by hours added to their working days by women entering the labor market.

Interest in Flexitime

Interest in flexitime has surfaced as part of a much larger discussion in the mid–twentieth century in Europe and the United States about the quality of work life and the role work plays in the whole of people's lives. From negative assessments about the nature and effects of work, and about current conditions and terms of work, at least three intertwined lines of inquiry and reform have emerged: to humanize working conditions, to accommodate workers' schedules, to facilitate workers' childrearing responsibilities.

To humanize working conditions. Because of the negative portrayals of blue-collar work, in particular, in the national investigations of work and the quality of life in the 1970s, a number of public and private organizations began investigations and programs to improve the quality of work experience, especially for workers in large impersonal organizations. (For example, O'Toole, funded by HEW, 1974; Andrews and Withey 1976; Campbell, Converse, and Rodgers, funded by the Department of Labor, 1976; Best and Stein, for the Department of Labor and The National Employment Commission, 1977; and Evans, for the Organization for Economic Cooperation and Development, 1973.)

Generally, the reform groups seek to involve labor and management together in determining what procedural and substantive changes would improve their work lives. Among such groups in the Washington, D.C., area are: the American Center for the Quality of Working Life; the Harvard Project on Technology, Work, and Character; and the National Council for Alternative Work Patterns, Inc. Substantively, these reformers emphasize the need of workers in bureaucracies to feel appreciated and trusted (e.g., La Bier 1980).

The ideal of flexible work schedules also represents an effort to extend to non-professional workers some of the same control over their time that professionals in and out of federal service have always had. Flexitime may be seen primarily as a new advantage for the 60 to 70 percent of federal employees who do not hold professional level jobs.[1] The changes permitted by flexitime for non-professional workers may also have related advantages for professionals by removing a major inequity and

[1]Similarly, current federal annual leave provisions allow employees to use leave days for virtually any reasons they choose, up to the total number allowed (depending on years of service). The purposes of annual leave are: first, to allow every employee an annual vacation period of extended leave for rest and recreation, and second, to provide periods of time off for personal and emergency purposes. Following a list of examples of such purposes (e.g., death in the family, religious observances, conference attendance, securing a driver permit), the regulation states: "These situations are not all inclusive, but are examples only, of the purposes or the kinds of absences for which annual leave is approved." In other words, the law permits people to miss work in order to stay home with sick children, or go places with children—but not explicitly, and as a direct trade-off with vacation time (Office of Personnel Management, Federal Personnel Manual 1969, p. 630-17).

source of tension between the two groups. To some extent, flexitime may ease professionals' pressures, too, by giving them the "right" to oversleep, or to spend longer in the morning at home, or to go to a child's ball game on a work day, or whatever—as long as the time is put in eventually, in accordance with the flexitime system in the particular workplace.

To accommodate workers' schedules. As a result of changing demographic factors, various groups of workers have been identified by age, sex, and functional characteristics as needing different working arrangements, particularly in the scheduling of work. The groups most often mentioned include: women with families; single parents (usually mothers); students; older workers; the handicapped; and increasingly, men with young families (see, for example, the rationales in Cohen and Gadon 1978; and Poor 1973). One example of efforts to make new scheduling arrangements is New Ways to Work, a job-sharing project in Palo Alto, California; it has become a national clearinghouse for information on jobs in which two individuals hold one position in order to increase career part-time employment opportunities (Meier 1978).

To facilitate workers' child rearing responsibilities. Efforts to prevent work demands on parents from having negative consequences for children have arisen in conjunction with the increased labor force participation of American women. The concern is exhibited in requests for child care accommodations at workplaces and in communities, for extension of public school hours to provide after school child care, and for greater flexibility in working hours for parents so they can attend to their children's needs more easily (see, for example, Gerzon 1970; Keniston and the Carnegie Council, 1977).

Flexitime was originally conceived precisely to address a work-family conflict. It was developed to allow women with young children to enter the labor force—while still taking the primary responsibility for daily management of housework and child rearing. Christel Kamerer's 1965 theory of flexitime (originally "gliding time") was an economist's response to a labor shortage in West Germany.[2] Yet flexitime has usually been introduced by employers to combat workplace problems—absenteeism, tardiness, and so forth. It was first put into practice in 1967 to alleviate auto traffic congestion at the Ottobruun Research and Development Plant of the Messerschmitt-Boelkow-Blohm aerospace company in Munich, which had twelve thousand workers on flexible schedules in 1977 (Nathan 1977). In this way, our study of flexitime returns to the origin of the concept in order to assess how helpful the program is to women—and other family members.[3]

United States Flexitime

No national estimates of the number of American flexitime users are yet available, but the Bureau of the Census planned to collect these data as part of the April 1980 Current Population Survey. One projection from a survey of companies on flexitime estimated that, by 1978, 2.3 to 2.5 million workers in the private sector would be on flexitime (Nollen and Martin 1978). So far, flexitime has been particularly popular in service industries, where operations are relatively independent, in contrast to assembly line or production units. The first American company to utilize flexitime was Control Data Corporation (in its Minneapolis plant in 1972). Many banks and insurance companies have introduced flexitime, as have the Firestone Tire and Rubber Company and the Sun Oil Company of Philadelphia. Several manufacturing plants have also experimented with flexitime, including General Electric, Motorola, and Olivetti (Kuhne and Blair 1978, p. 41).

[2]Christel Kamerer, the German economist who wrote the original article outlining the concept of flexitime, is cited in numerous articles on flexitime, but we have not been able to locate a copy of the article.

[3]Among western European countries, the Swiss have the largest portion of their workforce on flexible schedules (35 percent of one million people); Germany has 25 percent of its workforce on flexible schedules (five million people) (Nathan 1977, p. 37).

Organized Labor's Position

Unlike other worker-oriented reform movements, organized labor has not been in the vanguard of the efforts to create alternate and flexible work schedules. On the contrary, the most prominent national unions have been among the few organized groups to testify against expanded flexitime programs, primarily on the grounds that the legislation allows suspension of the requirement that pay for time worked beyond eight hours per day must be at overtime rates.

Women's interest groups tend to explain union opposition to this innovation on the basis of the fact that national union leadership is primarily male, and that males see less value in flexitime than do employed women who carry primary responsibility for direct care of their families. Consequently, where local unions have largely female membership, flexible scheduling tends to be supported. The Communications Workers of America, AFL-CIO, with a large proportion of female members, has been one of the few national unions to take a positive position on alternate work schedules (Communications Workers 1978).

To date, in both Europe and the United States, management, more often than labor, has taken the initiative in flexible schedule reforms (see Stein, Cohen, and Gadon 1976; Martin and Hartley 1975; Nathan 1977; Kistler 1977; Fiss 1977; Elbing, Gadon, and Gordon 1974; *Business Week* 1972; *Business Week* 1973; *Personnel Journal* 1975; Fellows 1971; and Golembiewski et al. 1975).

While employees on flexitime have been almost universally enthusiastic about the option, according to management surveys, about three-fourths of private sector flexitime programs are in non-unionized sectors of large companies (like Control Data and Hewlett-Packard). On the other hand, it is estimated that perhaps 50 percent of federal employees on flexitime are in bargaining units represented by unions (Keppler 1979; Cowley 1979).

The Conceptual Framework for This Study

Despite the expanding use of flexitime, and the claims for its benefits, almost no empirical studies have tried to isolate work schedules as a variable in order to examine their effects on families, as the present study attempts to do. Winett's longitudinal studies (1978) of small groups of families on flexitime are the only other current research focusing directly on work schedules' effects on families. Magnusson and Nilsson (1979) have done a broader study of the social effects of work hours and schedules in Sweden.

To investigate the effects of flexible work schedules on families, our research uses two conceptual models from the still-emerging theoretical approaches to the study of the family. The first of these is the *ecological* approach adopted by the Family Impact Seminar from Bronfenbrenner's theory (1977b, 1979).[4] This point of view assumes that human beings must be understood in social environments (as opposed to simply intrapsychically, for example). In this sense, families are one unit among many institutions in the environment—such as schools, health facilities, neighborhoods, churches, stores, transportation, and workplaces.

In relation to the central question for this study, Bronfenbrenner's ecological framework requires an analysis of the interactions between one small system, the family, and one larger system, the world of work. (Together these systems are shaped by the overarching institutional patterns of the *macrosystem* of the culture in which they exist.) The research design for the survey deals with only one feature of the work system, the schedule, and its impact on several dimensions of the family system. The second phase of the study—the follow-up interviews with a few survey respondents and their spouses—investigates a wider

[4]Bronfenbrenner's theory is derived from Lewin's topological constructs (1935, 1936, 1948, 1951). He also acknowledges the influence of Barker (1965, 1968) on his thinking. See Appendix B for a full description of the four structures, or systems, in Bronfenbrenner's ecological theory.

range of work influences on family life, in addition to schedules, and gathers information from more than one family member in a small group of families (see Chapter 7).

A second approach to analyzing the family is *family systems theory* as developed by mental health practitioners on the basis of clinical observations and therapeutic experiments (see, for example, Kantor and Lehr 1975; Haley 1976; and Minuchin 1975). Family systems theorists see families as groups of individuals connected to one another so that the roles, behavior, and interactions of each member affect other members. From a systems viewpoint, understanding of family life requires attention to what goes on between people, as well as to what occurs within the family. (This approach is especially relevant to the consideration of "equity" in family work in this study; see the discussion following.) Various researchers and family therapists who use a systems framework often stop short of acknowledging that the families themselves are embedded in complex social systems. Using an intra-family systems approach plus a larger ecological context ideally enables researchers (or clinicians or policymakers) to examine the connections between people within their families and within their larger social-environmental settings.

In order to single out what the effects of work scheduling, per se, might be on family functioning—from the expectations expressed most fully in the congressional hearings on flexible schedules—this study investigates whether people who have a choice of flexible work hours, in contrast to those without choice, report one of the following results: feeling less stress; spending more time in family work; doing a larger share of family work.

A number of other factors besides work schedules affect these aspects of family lives. Thus, as intervening variables, the design also takes account of the following differences among respondents:

- Whether they live alone or with others.
- Whether one or both spouses is employed (family earning

structure) for respondents who are married and living with their spouses.

- The ages and number of children, if any (family life cycle stages).
- The occupational level.
- The number of hours worked and commuted.
- Whether or not people have outside help with home responsibilities.

Many other possible influences on family functioning are not controlled for in the design, both because of measurement difficulties and on the assumption that the variations would tend to be proportionately distributed across the experimental and control groups. Factors not controlled for include: health; marital satisfaction; quality, cost, and ease of child care arrangements; family income level; availability of friends, relatives, and other support services; nature of the neighborhood; and job satisfaction. (See further discussion of the limitations of the theoretical framework and the research design in the Afterword.)

The Variables

The remainder of this chapter discusses each of the three dependent variables—family stress, family work, and equity in family work; the independent variable—schedule; and each of the control variables.

The Dependent Variables

Although many flexitime advocates claim that work schedule flexibility benefits families, few discussions of the topic specify just what flexitime is expected to change for families—or how benefits can be observed. Perhaps the most difficult aspects of attempting family impact analysis of this policy or any other, are: first, identifying which aspects of family life may be affected by the policy; and second, devising ways to measure how these factors are affected. For example, if the goal of a family impact analysis is to determine how a policy affects a commonly used but essentially abstract concept like *nurturing*, it is necessary to establish, for purposes of the analysis, what nurturing is, and how it can be observed—in order to see if it is affected negatively or positively by a policy.

Family Stress

As indicated above, one of the dominant themes of the family-related arguments in favor of flexible work schedules is that the flexibility is expected to help people, especially women, manage both job and family responsibilities more easily. Flexitime is expected to relieve the double burden carried by women employed full-time, by relaxing the requirements about which forty hours per week they have to be at work.

Both scholarly and popular writings in recent years have characterized concatenations of work and family pressures in contemporary life in terms of stress (e.g., Bebbington 1973; Booth 1977). Haldi (1977, p. 21) estimated that because more women are in the labor force, married couples in 1975 worked and commuted almost as many hours per week as their counterparts in 1900 (73 v. 75 hours), whereas in 1940 the average was only sixty-two hours per week. Headlines of recent newspaper and magazine articles illustrate the journalistic sense of the problems related to work, women, and family life:

> *The Washington Post*, April 9, 1979: "The Painful Questions: Stress of Work v. Family Decisions Is the Area's Major Mental Health Problem."
>
> Harden 1979

> *The New York Times*, October 20, 1976: "Family Stress Called a Menace to Health."
>
> Hicks 1976

> *The New York Times*, November 29, 1979: "Fast Changes in Society Traced to the Rise of Working Women."
>
> Dullea 1978

> *Ms. Magazine*, May 1979: "Is Success Dangerous to Your Health? The Myths—and facts—about Women and Stress."
>
> Ehrenreich 1979

Most of the research on family-related stress of the last fifty years (e.g., that inventoried by Burr 1973) utilizes the idea of a *stressor event* that precipitates change in a family, for example, disasters (e.g., Hill and Hansen 1960) or recent life changes like births, deaths, illness, or divorce (e.g., Rahe 1974; Dohrenwend 1973). This body of research looks at what makes families vulnerable to the stressor events and what strengths families show in coping with the resulting problems. Various theories have also attempted to classify types of stressor events in relation to various kinds of external stimulae and physical effects. As Hansen and Johnson (1979) point out in their consideration of definitions in family stress theory, while recognition of such stressor events has the virtue of taking into account the physical, phenomenological, and ecological contexts of an individual's stress, it omits considerations of stress as a process. However, even researchers more oriented toward the process of how people cope with stress still tend to conceptualize stress as resulting from change precipitated by particular causes or events (e.g., Datan 1975; Mechanic 1974, 1975; Lauer and Lauer 1976; Elder 1979; Zablocki 1976; Holmes and Rahe 1967; Selye 1956).

Several newer, less-developed approaches to studying family stress, which take external contexts into account, have begun to focus on "chronic," everyday stress in families (e.g., Croog 1970; Frankenhaeser and Gardell 1976; Piotrkowski 1979b; Stress and Families Project 1979; Ilfeld 1977). Within the ecological framework for this research, two particular theories of chronic stress offer useful ways of conceptualizing family stress as related to work schedules. The theories of Bronfenbrenner (1977b) and Pearlin (Pearlin and Schooler 1978) aim to understand stress in terms of the routine contexts in which it is experienced by individuals. In short, their theories address: first, the frustration and discomfort which normal people (not patient populations) encounter in their everyday lives; and second, the impact of social structures (like work) in contributing to these discomforts, or stress.

Both Bronfenbrenner and Pearlin define stress in subjective and objective dimensions. For Pearlin, feelings of emotional upset are *stress;* the conditions which cause stress are *strain.* Bronfenbrenner makes a similar dichotomy with different nomenclature. For him, feelings of frustration or dissatisfaction are *subjective, or phenomenological, stress;* the events, situations, groups, or individuals identified as the source of difficulty, discouragement, or dissatisfaction in a person's life are *objective stress*. For purposes of this research, stress is characterized as a negative factor in family life; therefore, it is assumed that to the extent flexitime may ease certain kinds of chronic stress, this would be a good result for families. But clearly some kinds and degrees of stress may also be advantageous, for example, in terms of helping people develop strength and skills to cope with change and problems.

In their own empirical research Bronfenbrenner and Pearlin consider stress in relation to particular social roles. Pearlin investigates how adults cope with stress as workers, as parents, and as spouses. Bronfenbrenner focuses on parents, and on the ways they feel stressed or supported by the circumstances in their work, their neighborhoods, and so forth—that is, in terms of the "ecology" of their "whole" lives.

The present study combines Pearlin and Bronfenbrenner's subjective and objective dimensions of stress, defining *stress* as: the experience of discomfort, pressure, tension, or frustration that may arise as people function in both their job and family worlds. But the emphasis here also differs slightly from the conceptions of Pearlin and Bronfenbrenner. Whereas they look at arenas of stress separately—that is, either in parenting or in working—this research deals less with how each role by itself affects people's well-being than with how the points at which the roles connect or overlap with each other produce stress for the individual in his or her family life. It does not deal with the individual's entire experience as a worker (e.g., involvement, boredom, pressure, challenge), but only with the way in which *timing* of the work, and *choice* of timing, may contribute to or ease the stress that person feels in also fulfulling a family role.

Marks' (1977) critique of the "scarcity" approach to time and energy challenges the assumption in this formulation that time pressures always have negative effects on people. He argues that multiple roles do not necessarily drain energy, but may increase it, depending on an individual's levels of commitment to particular activities. But for the purposes of our survey, the stress measures are based on the simpler, linear assumption that multiple roles in work and the family increase stress (which is not necessarily the same as "draining energy"). Ideally, future research on multiple roles and time pressures will develop more complex conceptual models that can better account for this complicated dynamic.

To examine how timing—or schedule—may affect stress, our study distinguishes two types of discomforts or pressures that employed family members may feel. One type of stress consists of worries people may have about whether they adequately accomplish everything they feel obligated to do in both work and family arenas (i.e., *role strain*). The other type of stress consists of assessments people make of how difficult it is to arrange their time to take care of particular personal or family responsibilities (i.e., *family management*).

Marks (p. 927) also argues that the "scarcity" approach to understanding the causes and effects of role strain confounds the distinctions among time, energy, and commitment that should be considered as conceptually different issues. This study focuses only on time in order to see whether or not the manipulation of this variable—by means of more flexible work schedules—affects people's feelings of stress related to their work and family activities and obligations, and whether or not it affects what they do in their family life. (See the following section on "family work"—a term suggested by Joseph Pleck to encompass both child rearing and home chores.)

Flexibility in work schedules could also be examined from the opposite side, namely, whether the absence of external time demands or structures could be as stressful for those who function best when their use of time is ordered by others as rigid scheduling is for those who do not. The fact of having more choice about the use of time might also affect personal inter-relationships negatively; for example, a wife might increase pressure on her husband to participate more in family life if she knew he had flexitime—and he might feel resentful. Again, more developed conceptual models may be able to take these issues into account in future research.

In the absence of existing measures to assess the kinds of stress described above, two new scales were developed for our study. The Job-Family Role Strain Scale taps internalized emotions, such as self-doubt, guilt, and pressure, in regard to the felt obligations about job and family. On a scale of 1 to 5, respondents indicated how often they felt the emotions expressed in a series of statements (for example, that they have a good balance between their job and family time, that they have to rush to get everything done each day, that time off from work did not match other family members' schedules well, that their job kept them away from their family too much).

The Job-Family Management Scale taps respondents' feelings about the logistics of family life for which they feel responsible. On a scale of 1 to 5, they indicate how easy or difficult they think it is for them to arrange their time to accomplish various chores, and errands, and to interact with other family members; for example, how easy it is for them to go to work a little later than usual, to adjust their work hours to take children places, to stay home when they are sick, or to do things with them. (See Appendix C for a full discussion of the conceptualization and construction of both scales.)

To the extent that the scales name specific circumstances in family life that employed members may find stressful, they respond, in some degree, to the stipulation by Hansen and Johnson (1979, p. 588) that the qualities of stressful situations that render those conditions problematic to individual family members be defined. Yet the specificity in the scales of areas in which people may feel stressed also undoubtedly distorts the reality of their routine stress since, as Piotrkowski (1979a) and others have emphasized, chronic stressors tend to be invisible. Or, as Elder (1979) has noted, even drastic change may be gradual and therefore unidentifiable.

Family Work

Child rearing. Beyond the predictions that work schedule flexibility will ease the logistical complexities of managing both the home and work—and thereby reduce stress—it is less clear exactly what flexitime supporters expect people to do differently, and when. The most general implication of their reasoning is that people on flexitime may be able to spend more time in family life, particularly more time with their children.

While debate about what kind of care is good or bad for children continues in current research and policy discussions, scholarly consensus at the moment seems to support the following general proposition: for optimal cognitive and emotional development, children must have a sustained relationship with one or more primary caretakers that includes both "enough" time and "quality" time (e.g., emotional and intellectual interaction between the child and adult). Summarizing recent research on which this conclusion is based, Clarke-Stewart reports:

> Children's development of social, emotional and intellectual abilities is greatly enhanced by the way their parents behave towards them. Early language development is perhaps most clearly influenced by parental behavior. . . . Although the time parent and child merely spend together or spend in routine caretaking is not related to the child's competence, stimulating

> mutual interaction of the pair most certainly is. . . . Intellectual development is also related to parents' attention and responsiveness to the child's needs, abilities and expressed desires. . . . Children's competence is related not to one or two single, isolated parental behaviors, but to a complex multivariate constellation or pattern of behavior.
>
> Clarke-Stewart 1977, pp. 58–60

In his time-budget reports, Robinson (1977a, 1977b) also reflects this view of the importance of parental time with children: "While parents may not consciously pattern their child care behavior as such, the time they devote to caring for and interacting with their children represents an investment that does determine how the child will mature and develop" (1977a, p. 69).

In addition, this sense of "what is good for children" is echoed by the "ecological" paradigm spelled out by the National Academy of Sciences:

> The problems of children are inseparable from the problems of those who are primarily responsible for their care and nurture. The characteristics of each child's environment are largely determined by the environment and circumstances of one or more adults, including, but not restricted to, the child's parents. It is therefore axiomatic that one cannot attempt to help children without influencing those adults, whether intentionally or not. This is especially true in the case of very young children because of the degree of their dependence and need for close psychological attachment to a particular adult.
>
> National Academy of Sciences 1976, p. 10

The assessment of the value of parents' time with children ideally requires evaluating the quality aspects of the interaction—mood, substance, and psychological dimensions. But consideration of what quality time may entail—and how much time is enough time—lies beyond the scope of this survey. Thus, the attempt here is to see simply whether parents on flexible schedules report that they spend more time, on the average, with their children than those on standard time.

In keeping with the emphasis in this study on the points at which family work and roles intersect, the amount of time people spend with their children is an *ecological variable* in the sense that time, and its use, constitute a central way in which people make the linkages among social institutions (or structures) in their lives. The focus on time as a dependent variable represents one aspect of the interaction between a small (micro) system, the family, and a larger (meso) system, the workplace (see Appendix B).

Given the a priori temporal constraints of employed parents—namely that they spend, on the average, at least a third of each twenty-four-hour period on their jobs and commuting—the reason for asking how much time they spend with their children follows from the implicit assumption (in current research) that the more time full-time employed parents spend with them the better. For purposes of this research, the *amount of time* is a proxy for the quality of the time. It is also implicit in the expectations for flexitime that if parents are less stressed, their time with the children will be of higher quality, as well as of longer duration.

Whether these more subjective benefits for children will flow from more time and less stress also will depend on many additional factors that individuals bring a priori to their circumstances. Personality and health issues are primary determinants, but perhaps no more so than environmental circumstances, such as the attitude and activity of the spouse, the nature of the community, friendship networks, and social comparisons and expectations for oneself. In addition, the nature of the interaction, whether it encourages children to develop both secure and comfortable feelings about themselves on the one hand, and lively cognitive interests and learning skills, on the other, will also constitute the critical "substance" of "quality time." (For a discussion of these "ecological" factors, see Bronfenbrenner and Cochran 1976.)

Home chores. Parental time with children, or child rearing, is often, perhaps usually, intertwined with other activities related to managing a household (illustrated in the vignette in Chapter 3). Yet those who recommend flexitime as beneficial to families with children do not spell out exactly what they expect flexitime to improve with respect to the household dimensions of family life. It is unclear whether women on flexitime are expected to be able to reduce their time on home chores; or whether it is supposed to give them more time at certain points in the day which they could use for their family (e.g., to commute, to get dinner, to interact with the children, or to shop). Some advocates imply that flexitime should reduce the time women spend on home chores because it will allow men to increase the time they spend on children and activities related to their care.

Given the ambiguity of the expectations on this issue, the measures of stress and time with children are supplemented by an exploratory investigation of whether flexitime seems to increase or decrease the amount of time people say they spend on home chores, with attention to the difference for men and women. For example, do men on flexitime spend more time on chores than man on standard time? Or vice versa for women? This variable is considered in the same ecological context as time with children, that is, time constitutes a link between the *microsystem* (the family) and the *mesosystem* (the workplace). Again the question for study is whether a change in the mesosystem (i.e., more choice in schedule) produces a change in the microsystem (i.e., more time in home chores).

Using "summary" measures devised by the Michigan Survey Research Center, respondents were asked to estimate how much time they spend each day on all home chores and all child rearing—separately. (In the analysis the measure was transformed to weekly hours.) While the goal was to provide preliminary information on whether minor schedule flexibility, of the sort available in the sample agency, affected the amount of time people spent on home chores, the summary measures are inevitably a crude means of assessing such phenomena. As time-budget researcher Robinson has emphasized, even for the painstaking time-diary approach (where people report their daily use of time in fifteen-minute segments): "Simple time expenditure data on housework do not reflect these subtle role demands that are involved in the performance of housework; that is, the constant attention required throughout the day, week, or year; the continual changes and decisions about scheduling and priorities; the monotony of the hectic and the menial" (Robinson 1977a, p. 62).

For a full discussion of the ways in which these domestic activities have been researched in the past, and of the way in which the summary measures are conceptualized and used, see Appendix D.

Family Equity

A third reason that flexitime has been promoted as beneficial to family life is to foster more equality for males and females in work and family roles. The implicit prediction is that greater work schedule flexibility may allow women to participate more easily in the job market, and men to participate more easily in family activities. The Rapoports outline a sympathetic context for understanding the current prominence of this concern, building on formulations by Safilios-Rothschild (1976), Young and Willmott (1973), and their own earlier research (1971):

> Social changes do not occur in a simple linear way. They occur in a piecemeal fashion—now in the workplace, now in the home—with different areas providing bottlenecks of further change at particular times. At the time of the first edition [of *Dual Career Families* 1971] it was assumed that the bottleneck in the movement towards greater equality between men and women was in the workplace and its discriminatory

> employment and advancement practices. . . . No doubt that the conditions of the workplace presented barriers and impediments to the general advance of women towards parity with men. . . . However, . . . there was [also] a domestic bottleneck that would impair women's capacity to respond to occupational opportunities even if all the prejudices and discriminatory practices were removed. This domestic bottleneck was formed by the tradition of men's and women's roles and the associated mythology underpinning them.
>
> Rapoport and Rapoport 1976, p. 354

> In the home, by contrast, the task of equalization has only begun. Resistance to this process seems to center on three linked issues: how domestic and child care work is to be accomplished; how to deal with the external occupation demands in relation to family requirements and values; and how to coordinate and resolve the apparent conflicts arising from these two sets of demands. . . . The first of these issues presents few organizational problems in the technical sense, but many emotional problems stemming from quite primitive levels of sex-role conceptualizations. The second issue centers more on the workplace. In the recent past, the workplace outside the home has been organized for men, more or less without reference to the needs of women or children.
>
> Rapoport, Rapoport, Strelitz, and Kew 1977, p. 19

As early as 1942, Bernard, among others, touched on this issue in terms of family problems arising from the lack of synchronization between the work of homemakers and the work of labor force participants (Bernard 1942, pp. 535–36). More recently, the problem has been seen in terms of cultural values and social structures. For example, Pleck (1977b) thinks that each sex is constrained from participating as fully as the other in the alternate arena largely as a result of sex role attitudes and traditions. He considers these male-female inequities in work and family as part of a "work-family role system" in which women's roles are limited by their family responsibilities and men's family roles are limited by their job responsibilities. (Also see Reinhold 1977, for a recent journalistic account of the issues.)

On the same issue, Lloyd (1975, p. 13) argues that children traditionally have been considered the main source of complementarity between husbands and wives in marriage, and that the presence of children has been a deterrent to the labor force participation of women and an inducement to the labor force participation of men. But economic and demographic changes (e.g., decline in fertility and growing equality in male and female training) have now combined to make men and women more substitutable for each other in household and market production.

Yet the different role expectations for each sex continue to be sustained by a variety of attitudes and institutional structures, including work settings and schedules, which remove employed people from their homes most of the daylight hours. Other institutional structures that reflect and sustain the different roles for men and women include: the unavailability of universal child care systems for preschool children, the unavailability of universal after-school care for school-age children, the lack of provision for employed parents to use sick leave for needs of children or other family members, the dearth of paternity leave policies among employers, and the related reluctance of men to use leave when it is available. Our study examines whether offering a little more choice in one of these structures—namely, work schedules—will influence attitudes and behavior towards more equity in male and female roles in the family.

In terms of the ecological theory underlying this research, again, the question is whether or not a change in the mesosystem (i.e., flexible schedules in the workplace) may in turn affect the microsystem (the family). This "equity" variable has also been considered in terms of *family systems theories* (e.g., Kantor and Lehr 1975; Haley, 1976; Minuchin, 1975). As discussed above, these theories postulate that if the behavior of one family member changes, the behavior of other members will change also (in some way). This expectation is based on the assumption that the family members are linked to one another in a dynamic chain in which an alteration in one link reverberates through the others. In this sense, family systems theories are parallel to the larger ecological theory of micro-, meso-, and macro-systems. Thus, in relation to work schedules, the hypothesis is that if one spouse changes his or her behavior, in this case, the amount of time spent on family work (as a result of more flexible work hours), then the proportion of time spent on family work by each spouse will decrease or increase accordingly. For example, if a husband on flexitime increases the amount of time he spends on family work (and the amount of time his wife spends stays constant), his share will increase, and the *equity* in family work will increase in that family (assuming the wife spent a larger amount of time on family work than he did in the first place).

Recent research has considered the consequences of unequal male and female roles in work and family in ways similar to those articulated in the congressional hearings. In various studies the effects of "asymmetrical" division of labor by sex, in jobs, and at home have been seen in relation to five major issues: (1) the heavier total workloads borne by women; (2) the family opportunity costs for men; (3) the employment opportunity costs for women; (4) the effects on children; and (5) the effects on marital satisfaction.

The limitations of this study permit only the first two of these issues to be considered. By collecting data on time spent on family work, the study gathers some information on whether families in which one or another spouse has a flexible schedule tend to share family work more equally, and thereby reduce the proportionately heavier female family work load and increase the male family work role (measured by time in family work). (Related aspects of the other three issues are discussed in Appendix E.)

Heavier total work loads for women. Despite variations in measurement categories and techniques, the reports of the major time-budget studies in Europe and the Americas in the 1960s and 1970s suggest that the amounts of daily time spent in home chores in different countries are remarkably similar—and that the amounts of time spent on these activities by males and females are dramatically different (see Meissner et al. 1975; Joseph Pleck 1976; Robinson 1977a and 1977b; Szalai 1972; Walker and Woods 1976).

Employed women everywhere spend an average of three to five hours a day on home chores; employed men, on the other hand, spend an average of one-half to one-and-one-half hours per day on such chores. Even though wives average fewer hours at work than husbands (about five-and-one-half versus almost eight hours), wives still work a total of two hours a day *more* than husbands, when time on family chores and time on employment are added together. On the other hand, full-time housewives and employed husbands have more nearly the same length workdays. (In different countries and studies, the time housewives spend on home chores ranges from six to ten hours per day.)

In explaining the reasons for the consistent amount of time spent in housework over different historical periods, and in spite of technological change, Myrdal and Klein (1956, p. 38) argue that labor-saving devices raised standards of housekeeping but did not reduce the time women spent on housework. Contrasting the time spent in home chores by employed and unemployed women, Myrdal and Klein also raise questions about whether the amount of time housewives spend on home chores is necessary, or whether such women simply expand their activity to fill the time in order to convince themselves and others that they are productive and that what they do is essential and time-consuming. They suggest that many employed people's work is "never done," too, in the same sense that they leave things unfinished and go home from their offices or factories; the difference is that they leave the work setting, whereas the housewives' unfinished work is ever-present and visible.

A contrary view would emphasize that *family work* often cannot simply be left unfinished at the end of a day because of the daily dependence of others on its being done—for example, care of young children, food shopping, meal preparation and dishwashing, laundering, and putting away clothes. The tasks of housework lend themselves to being done more often, or more thoroughly, or with more care and imagination—if the "doer" is so inspired. But the fact of the disparity between the time spent on family work by employed versus unemployed women leaves major unanswered questions that may be usefully explored by studying the nature of both women's and men's commitments to and rewards for certain kinds of work and the effects of these on energy (e.g., Marks 1977).

The expectations set forth in the congressional hearings and elsewhere are that if men had more flexibility in their work schedules, they might share more family work, and thereby help to alleviate employed wives' work overload. (Scanzoni 1979 calls this the "pro-social" or "altruistic" strategy to change men's roles.) The expectations for women are less clear, as indicated in the previous discussion on family work. For example, if women on flexitime increase their time on family work, the imbalances between males and females simply increase. At the same time, if employed women can better allocate their time for work and home responsibilities, they may feel less stress. Thus, our research first measures whether men and women on flexitime spend more or less time on family work than their counterparts on standard time. Then, looking at the proportionate amounts of time spent by husbands and wives in each family represented in the study, we determined whether family work is shared more equally in the flexitime or the standard time agency.

Family opportunity costs for men. Although few empirical

studies have explored the question of what men lose by their separation from daily family life—in developing the emotional and "nurturant" sides of their personalities and in relationships with their children, for example—a number of authors writing on sex roles in recent years have suggested that men would have healthier, more balanced lives if they were allowed and encouraged to spend more time in family life. Scanzoni (1979) calls these arguments for more equity in male-female roles the male "self-interest" strategy (see also Balswick and Peek 1971; Feigen-Fasteau 1974; Farrell 1974; Levine 1976; Pleck and Sawyer 1974; Joseph Pleck 1976; Liljestrom 1978; Oppenheimer 1974; Gronseth 1971; Tognoli 1979).

The dominant theme of all the writers on equity may be summed up by the following: "Two lines of development in men's roles have recently begun to unsettle this pattern [of a father's lack of involvement with daily family life:] dissatisfaction with the rewards available to them in a lifestyle so concentrated on work and its accompaniments . . . [and] awakenings in men of a desire for more involvement with their wives and children" (Rapoport et al. 1977, p. 241).

This question of what men miss because of the asymmetry in family and work life is not measured directly in this study but the expectation is that with more flexibility in the scheduling of work, men will feel freer, and indeed encouraged by their work structures, to spend more time in family life. However, reports on male use of family leave time in Sweden (now equally available to men and women) suggest that relatively few employers encourage men to use such leaves, and the percentage of men using the leaves is growing very slowly (Morgenthaler 1979).

Our survey assesses simply whether males on flexitime report spending more time in family work than males on standard time and, in turn, whether the flexitime families share family work more equally. The survey does not examine the forces, other than work schedules, that provide incentives, punishments, and rewards to men for staying more involved in work than in the family (see, for example, Marks 1977; Aldous, Osmond, and Hicks 1979); but these issues are taken up in the interviews and discussion in Chapter 8.[5]

Independent Variable

As is clear from the previous discussion, the independent variable for the study is *work schedule,* defined as "the regulations in a particular workplace regarding when employees must be at work." Within the theoretical framework for the study, the independent variable is part of the mesosystem (work) and the dependent variables—stress, family work, and family equity—are part of the microsystem (family).

Standard schedules are those arrangements requiring all workers in a particular federal agency, or office, to be at work during the same hours without individual variations (e.g., 9:00 A.M. to 5:30 P.M.). Flexible schedules in this study refer to those arrangements in which workers may exercise some individual choice about the time of arrival and departure, as long as they work a full eight hours per day.

The sample agency in this research offers what is called a *modified flexitime program:* workers may select a routine arrival time between 7:00 and 9:30 A.M. and must generally stick to this plan with a leeway of fifteen minutes or so each day; they can change their plans at any time with two weeks' written notice to their supervisors. The mid-day lunch break and afternoon leaving time also are scheduled by worker choice (between 11:30 and 1:30 at lunch, and 3:30 and 6:00 in the afternoon). The "core" hours when all workers must be in the agency are from 9:30 to 11:30 and from 1:30 to 3:30. In practice, some offices are more lenient than others in permitting variations of the flexitime program, so that, for example, people may vary their starting, lunch, or leaving times on a daily basis, as necessary. Because

[5]See Appendix F for a discussion of the correlations between the dependent variables.

most people who use flexitime tend to "flex" early, each office tends to work out its own ways of rotating responsibility for staying late in the afternoon so that the office is "covered" during these hours. (Each office must have someone present from 8:30 to 5:00.) Further, if a conflict occurs between work requirements and an employee's schedule preference, the work obligations prevail and the supervisor makes the relevant decision.

Because the flexitime program considered in this study represents a modest variant in the range of possible kinds of work flexibility, any differences that appear between the two groups might be expected to be larger if the experimental group had more flexibility (for example, longer flexible bands, a system for averaging time over an entire month, or *flexiplace*, enabling people to work at home some days, and so forth). Also, although the standard time agency does not have a formal flexitime program, employees may have more or less flexibility depending on the nature of their work, or the time of the year, or the discretion of their supervisors. Overall, these kinds of variations presumably cancel each other out, given the similar nature of the work performed in the two agencies.

Consideration was given to breaking the flexitime group in two parts: those who reported that they had changed their hours since the agency began the flexitime program (a year ago), and those that did not report a change in hours. However, this dichotomy was not used on the grounds that even if people did not routinely alter their schedules, the fact of having the choice to do so might make a significant difference for the flexitime group in contrast to the control group, where the option is not available. Furthermore, separating out those who reported using the option program would involve a self-selection factor that might obscure the effects of the stimulus. Thus, operationally, the independent variable was determined according to which agency the respondent worked in.

From the perspectives of ecological and family systems theories, the expectations are that different variants of the independent variable, namely, different schedules, will have different effects on the dependent variables, namely, the three aspects of family life selected as outcome measures.

Control Variables

To determine whether the respondents in the two agencies differed on the three dependent variables, the design was constructed to test for differences of means (1) on the stress scores, (2) on the hours spent on family work, and (3) on equity in the time spent on family work between spouses. After the initial difference of means test—before taking into account other possible influences on the dependent variables—the means of the flexitime and standard time groups were compared again, controlling for six additional variables. (The control variables chosen for the multivariate analyses derive from the ecological and family systems frameworks for the study, as discussed below.)

Family Earning Structure

This variable refers to both the sex of the respondent and whether or not the respondent is in a one- or two-job family.[6] In terms of both ecological and family systems theory, each of these factors, plus their combination, should bear on how helpful flexitime may be.

As discussed in the foregoing section, it is clear that men and women have had very different employment and family work patterns. Thus, the effects of the mesosystem (work) on the microsystem (family) will tend to differ for males and females. Women who are employed full-time spend slightly fewer hours per week on the job, on the average, than men employed full-time (Robinson et al. 1972, p. 118). But all women—employed and not employed—spend many more hours per week on family work than employed men, on the average (e.g., Rallings and

[6] This way of identifying subjects in this study was suggested by J. Pleck; see his paper on job-family interference (Pleck et al. Sept. 1978); he uses the labels "employed husbands, wife employed" and "employed wives, husband employed" for the same purpose.

Nye 1979, p. 205). Thus, employed women's total work load exceeds men's, in terms of hours per week. For this reason, the sex of the respondent was controlled for on the assumption that all women in the sample would report more time on family work than men, regardless of schedule.

In addition, it was anticipated that workers whose spouses were not employed would spend less time on family work than those whose spouses were employed. Most of the respondents in the sample whose spouses were not employed were male. Thus, it was expected that any differences in total hours on family work for men probably would be influenced by whether or not their wives were employed—in addition to any differences related to schedule.

Family Life Cycle Stage

This variable has been defined in many different ways in the literature (e.g., Aldous 1978; and Glick 1977.) In general, *family life cycle stage* refers to the characteristic stages of family life development, for example, newly married couples, families with young children, families with children in transition to adulthood, families with no young dependents, and families with elderly dependents.

For our study, families "with children" are defined as those having children under eighteen years of age living at home. Respondents with children eighteen years of age or older living at home were included in the "no children" category, under the assumption that children in this age group typically require less parental time than younger children, and indeed often contribute more time than they take; or they no longer live in the same household with their parents.

Both time-budget data and child development studies suggest that the age of the youngest child and the total number of children at home are the two most important factors in determining how much time parents—usually mothers—spend in family work. Preschoolers require the most parental time. Parents with teenage and grown children spend half as much time on family work as those with babies under a year. Adults with no children spend the least time (see Walker and Woods 1976, p. 33). Again family systems theory applies: the presence and activity of some family members (children) affect the behavior of others.

The family life cycle stage variable was divided into three categories: respondents with no children, respondents whose youngest child is under six, and respondents whose youngest child is six to eighteen years old. Our expectations were that parents with children under six spend the largest amounts of time in family work irrespective of work schedule and that the more children people have, the more time they spend on family work. Therefore, both the number and age of children were controlled for in order to see what effects different work schedules may have on the dependent variables.

Occupational Level

This variable constitutes an ecological factor, within the guiding theory of this research, in the sense that by controlling for occupational level the design takes account of the effects of another mesosystem factor (like schedule) on aspects of the microsystem (i.e., family stress and family time). *Occupational level* is defined as a rank in an employment hierarchy.

For reasons discussed earlier in this chapter, it was expected that employees at higher government service (GS) grades would tend to work longer hours than those at lower GS grade scale levels, and that, therefore, they would be less inclined, required, or able to use flexitime in the family-related ways measured in the study. For those higher GS grade employees overtime might explain some differences in amounts of stress and time, irrespective of schedule. Although the requirement is not printed in any government job description, aspiring and/or high-level professionals often expect to work routinely for more than forty hours per week, with increasingly heavy work loads, as a necessary step towards promotion to higher levels of responsibility and recognition.

High level government professionals work long hours largely because the ethic of hard work is nurtured in them throughout their formative years and during educational training. In addition, the cultural expectations at the top of most professions make work, for its own sake, the cause to which most waking hours and effort should be devoted. In the private sector, attorneys, physicians, academics, and executives exhibit the same pattern of working intensively at their respective enterprises. In turn, these hard-driving professionals judge those who work with them by the standards to which they hold themselves. For such professionals, the years of proving oneself and of striving for greater opportunities often coincide with the years of bearing and rearing children.

Competitive incentives are doubtlessly mitigated to some degree by the Civil Service tenure system, which tends to keep and promote people through the ranks whether or not they work longer hours. (See further discussion of these issues in the interviews, Chapter 8.) Furthermore, since higher GS grades usually coincide with higher socioeconomic status than lower GS grades, especially in one-earner families, it seemed possible that high-level GS workers might have more outside help, thereby reducing the total hours of family work—home chores in particular.

On the other hand, for non-professional workers (defined below), who are covered by the Fair Labor Standards Act, the average amount of overtime recorded and paid in Fiscal Year (FY) 1976 was only twelve minutes per day per federal worker, thus suggesting that for those at the lower half of the occupational scale overtime itself does not preclude taking advantage of flexitime—for family or other reasons.

But the average overtime figure is not very useful, since most overtime is worked in "production" units (e.g., clerical, mail, and computer operations) and by high-level professionals. "Volunteer" overtime technically does not exist since supervisors must request and receive approval for overtime in advance.

Generally, employees who are exempt from the provisions of

the Fair Labor Standards Act (approximately those above GS 10, that is, 40 to 50 percent of the GS labor force) seldom receive overtime pay, although they are eligible up to a total annual earnings of $47,025; that is, a mid- or senior-grade GS employee could work overtime and get paid up to a total annual income of $47,025. More often, they take *compensatory time* in exchange for longer hours worked (which is given at the rate of one hour off for each hour worked, not at time-and-a-half, as with money).

One of the reasons few professionals from GS 9 and up get overtime pay is that the arrangements must be made and approved by supervisors up the line in advance. In addition to the difficulty in foreseeing the need for overtime in professional work, the mechanism for approval is complex (e.g., it requires justification in writing), and, therefore, is a nuisance to request. Finally, requests for professional overtime are seldom approved because of the strong expectation that professionals should be able to complete their assignments within regular hours—or if not, they should get the work done anyway (Cowley 1977; Dise 1977; Gavan 1977).

To measure occupational level, the study design uses a combination of the respondents' GS grade (3–18) and their type of work, as defined by the Civil Service Commission, PATCO (professional, administrative, technical, clerical, and other) categories. All subjects were grouped into two occupational levels, high and low. The first group (high) includes all those classified as professional and administrative employees at GS level 11–18; the second group (low) includes all respondents classified as clerical, plus any respondents in professional, administrative, and technical categories at grade 10 or below (see Appendix G, Government Service Job Classifications, PATCO).

Total Hours of Work

This intervening variable was expected to have effects similar to occupational level, and to correlate highly with it. We anticipated that long hours of work would tend to preclude employees from being able to use the kind of schedule flexibility available in the sample flexitime agency for family or any other non-work purposes. In terms of the conceptual framework for the study, another mesosystem factor (total length of working hours) was expected to have effects on the microsystem (the family).

Although forty hours of work a week is the required norm, a small percentage (2 percent) of federal employees work *permanent part-time* (i.e., less than forty hours per week), and another minority work more than forty hours per week, for the reasons suggested above in the discussion of occupational level. Thus, the expectation was that a range of hours worked would be reported, with the majority clustered at forty hours per week. Because those workers at higher GS levels are predominantly male, it was anticipated that more males than females would report longer hours of work, regardless of whether they were on flexitime or standard schedules.

Time Commuting

In addition to variations in total work time, it was anticipated that the amount of time people spent travelling from their homes to their jobs and back would differentially affect their stress and amount of time on family work. In theoretical terms, another mesosystem factor—work transportation options—would influence the microsystem. Thus, the amount of time commuted is controlled for in the multivariate analyses.

Outside Help

This variable was conceived in family systems theory terms; that is, people who paid someone other than family members to help with child care and housework were expected to spend less time on these activities themselves, and/or feel less stress in managing both their jobs and family work than those without "outside help." Thus, the design controlled for whether respondents had any outside paid help with family work.

Hypotheses

As indicated in the preceding discussion of the variables, there are three central hypotheses for the study. They are that workers on flexitime, in contrast to those on standard time, will: (1) feel less stress; (2) spend more time on family work (home chores and child rearing); and (3) do a larger share of family work (in their families).

Each of these hypotheses is also tested for various subgroups of the sample that are of particular interest, in light of the expectations for flexitime. The subgroups considered are: husband and wife families; single adult families; parents and non-parents; people with employed spouses; and people with non-employed spouses. Similarly, time on family work and equity in family work are compared for the above groups for which these variables were relevant. (See Figure 5 at end of Chapter 5 for a depiction of the subgroups and dependent variables considered for each group.)

5

The Survey Methodology

THE objective of this experiment in family impact analysis was to determine if a particular work policy, namely flexitime, can help workers to balance their jobs and their family lives. The aspects of family life selected for the measurement of the effect of flexitime were family stress, family work, and equity of spouse family roles.

In terms of the theoretical framework for the study, this survey is an "ecological experiment," in Bronfenbrenner's sense. It uses an "experiment of nature" (i.e., in this case an existing work policy) with heuristic purpose; that is, it investigates the accommodation between persons and their milieu by systematically comparing the effects of a structural component (schedule) and one environmental system (the government agencies) on another environmental system (workers' family lives), while attempting to control for other sources of influence.

The Sampling Design

The optimal design for assessing the effects of such a program would be an experimental one in which subjects are randomly assigned to treatment and control groups, and the independent variables can be manipulated so that their effects on the dependent variables can be observed (Campbell and Stanley 1963;

Warwick and Lininger 1975). However, the nature of the federal government and its flexitime programs did not permit direct control of a flexitime experiment in this way.

Alternatively, our first hope was to observe a group of workers before and after the introduction of a flexitime program, and then to compare their responses on selected measures (i.e., a pre- and post-hoc plan) as Winett (1978) had done in his exploratory studies of flexitime effects on parents with young children in two federal agencies. We also located two federal Washington-based agencies with appropriate populations and flexitime timetables for a pre- and post-hoc experimental design, but labor union negotiations within both agencies ultimately made management reluctant to permit the study at the time.

Thus, the reality of the research environment, and the goal of the Family Impact Seminar to complete the study in a year, required using a comparative ex post facto design. Such a design has more limitations than a true experimental design; namely, that if differences between the two groups appear in the measures, they may result from factors other than the utilization of flexitime by one of the agencies (Kerlinger 1973).

However, Bronfenbrenner has argued that "there are instances when a design exploiting an experiment of nature provides a more critical contrast, insures greater objectivity, and permits more precise and theoretically significant inferences than the best possible contrived experiment addressed to the same research question" (Bronfenbrenner 1979, p. 36). Perhaps this study, with its design limitations, is a case in point. (This possibility is discussed further in Chapter 9.)

To compensate for this design limitation, the agencies were matched as closely as possible on a number of criteria that otherwise might have accounted for differences between the two groups on the family measures. To facilitate this "matching," a purposive sampling procedure was used to select the agencies instead of, for example, randomly selecting subjects from all the Washington area agencies with flexitime programs. Agencies were picked that were as similar as possible on the following characteristics:

- Proximity: their locations were to be near enough to each other to equalize the commuting factors, given the likelihood that differences in commuting distances—and means of transportation—could directly affect the degree to which flexitime might help workers' family lives.
- Sex ratios: they were to have similar proportions of male and female employees to preclude differences in employee experiences and expectations about the respective working environments.
- Job grade distribution: they were to have similar ranges and proportions of Government Service (GS) grades for men and women, to assure comparability of level of work expectations.
- Agency mission: they were to have similar purposes and activities so that the process of daily work would be relatively comparable.
- Union activity: they were to have similar experiences and expectations with respect to unionization, so that one group would not be more or less concerned about this issue than the other.
- Race: they were to have similar proportions of racial minority employees at comparable grades to preclude variations in levels of "affirmative action" or interaction between races.

A two-stage procedure was used to select the agencies and the subjects within each agency.

Selection of the Agencies

Finding agencies to meet these criteria was more complex and time-consuming than anticipated. First, no central office in the federal government keeps information by agency on all the criteria listed above. Moreover, within each agency, no one office has all the relevant data; for example, the personnel office may have the GS grade and job code information, while the equal employment opportunity office may keep the sex and race data. Nor is the Office of Personnel Management required to know of all the agencies, divisions, or bureaus that have instituted flexitime. Finally, non-quantifiable issues like unionization tensions, differences in supervisor styles, issues of subtle coercion, and parking criteria (e.g., that only cars carrying four passengers can park in the department garage, thereby tying workers' schedules to those of their carpool mates) represented important differences that could be evaluated only through visits to the agencies and informal conversations. Some flexitime agencies had to be eliminated when there was no nearby standard time agency, or when union contract negotiations were under way. Before settling on the two agencies that participated in the study, preliminary discussions, visits and/or planning sessions were held with sixteen different agencies, as well as information-gathering conversations with an additional dozen.

The two agencies selected for the study were the Maritime Administration (MarAd), which had been on flexitime for one year at the time the survey was conducted; and the Economic Development Administration (EDA), which was on standard time, and had no plans to introduce flexitime.

The two met the preceding selection criteria in the following ways:

- Re proximity: the headquarters and vast majority of employees of both agencies are located in the main Commerce Department Building, which is a large downtown Washington office building (see the description and photographs in Chapter 6).
- Re sex ratios: the agencies had very similar proportions of men and women employees—the standard time agency was 54 percent male and 46 percent female; the flexitime agency was 58 percent male and 42 percent female.
- Re job grade distribution: the average male job grade in both

agencies was similar (standard time = 12.7; flexitime = 12.2). For women the average grade was slightly higher for standard time women (standard time = 8.2; flexitime = 7.4).

- Re agency mission: both are grant and loan making agencies.
- Re unionization: neither agency is unionized nor were organizers active at the time of the study.
- Re race: both agencies had close to one-third minority and two-thirds white employees (standard time, white = 66 percent, minority = 34 percent; flexitime, white = 73 percent, minority = 27 percent).

Thus, the agencies were well-matched in terms of the above criteria with the exceptions mentioned, plus one other difference: the flexitime agency is twice as large as the standard time agency. This difference was compensated for in the subject selection procedure. (See Chapter 6 for a description of the settings and functions of the two agencies and how flexitime was started in the Maritime Administration.)

Selection of the Subjects

In the case of the smaller, standard time agency, all 413 employees were included in the study. In the flexitime agency, initially a 50 percent random sample was drawn from a total of 813 employees; thus the original random sample was 406. Because the flexitime agency had proportionately fewer women at high GS grades than the standard time agency, the initial flexitime sample was increased to include thirty additional women at GS 11 and above, in order to facilitate analyses of the responses from women in professional levels.

Similarly, when it was discovered, as the questionnaires were returned, that the original sample of respondents included too few parents of children under eighteen to permit multivariate analyses in particular subgroups (e.g., single mothers and two-earner families), all flexitime employees whose names had not been drawn in the original sample were telephoned; they were asked if they had children under eighteen years of age, and if they did, were invited to fill out a questionnaire. One hundred and six more questionnaires were distributed to this group. To compensate for the distortions introduced by the two instances of oversampling, and to take advantage of the larger sample size, a weighting formula was used throughout the analysis.[1]

Background Characteristics of the Sample

The groups of employees in the two agencies were well-matched on most of the criteria related to employee characteristics. They were similar in all of the following respects: proportions of men and women (standard time: men = 51%; women = 49%; flexitime: men = 55%; women = 45%); whites and minorities (standard time: whites = 67%; minorities = 33%; flexitime: whites = 70%; minorities = 30%); educational levels, commuting time, overall job satisfaction and involvement, and average Government Service grades (see Table 1).

However, there were significant differences between respondents in the two agencies on three work characteristics: standard time personnel worked an average of two hours per week more than flexitime employees ($p \leq .001$); the average GS grade for flexitime women was lower than for standard time women (GS 6 versus GS 10, $p \leq .05$); and flexitime employees were much more satisfied with their work schedules than standard time workers. Almost one-third of the flexitime people

[1]The resulting sample then resembles a traditional stratified sample; therefore the standard methods for calculating means and variances were applied. The weighting formula to compensate for oversampling (of high-level women and parents in the flexitime agency) was developed by taking the sample numbers for each subgroup (defined by sex, occupational level, and parental status) and dividing them by the population numbers for each subgroup (as provided by the agencies). These results were used as the weighting factors to adjust the sample cases to the proportions in the total population in each subgroup (Kish 1965).

TABLE 1
COMPARATIVE BACKGROUND CHARACTERISTICS OF RESPONDENTS IN THE TWO AGENCIES (INFORMATION COLLECTED IN THE QUESTIONNAIRE)

Characteristics	*Standard Time*	*Flexitime*
Work-related		
Percent completed 4 years college	57.3	52.6
Average job grade—total sample (range is 1–18)	11.5	9.2
Average job grade males	13.0	12.8
Average job grade females*	10.2	6.5
Average hours worked per week	43	41†
Average one-way commute (in minutes)	45	45
Overall job satisfaction (on a 1–5 scale)	3.6	3.8
Job involvement (on a 1–5 scale)	3	3
Percent very satisfied with schedule	15.2	31.0†
Family		
Household composition:		
Average number of persons in household	2.8	2.9
Percent with relatives living in household	9.9	18.6†
Marital status:		
Percent married	62.0	58.4
Percent separated, divorced, widowed	16.2	12.7
Percent single	21.8	29.0
Parental status among total sample:		
Percent no children	34.4	40.8
Percent with children (all ages)	65.6	59.2
Percent with children under 6 years at home	17.6	15.9
Percent with children 6–18 years old at home	25.9	24.2
Percent with children over 18	22.0	19.0

*The mean female job grades reported by the agencies were standard time = 8.2, flexitime = 7.4; the mean male grades reported by the agencies were virtually the same as the sample means.
†Significant at $p \leq .05$.

were very satisfied with their schedules while only 15 percent of the standard time people were very satisfied with their schedules, $p \leq 000$ (see Table 1).

For family characteristics, again the two agencies were similar in most respects, for example, in terms of proportions married, or with children (see Table 1). The only significant difference between the agencies on family characteristics was that 19 percent of flexitime workers had other relatives living in the household, in contrast to only 10 percent of standard time workers ($p \leq .05$).

Measurement

Given the nature of the settings in which the data were to be collected and the time constraints of the Family Impact Seminar, it was decided that the most unobtrusive, acceptable, and cost-effective instrument to collect the information would be a brief, self-administered questionnaire that invited federal employees' interest and participation in the effort to see how policies of their workplace might affect their family lives.

As indicated above, the study attempted to assess the effect of schedule on three family factors (dependent variables). Family stress, defined as tension or pressure arising at the points where people's work and family roles connect or overlap, is measured by two scales created for the study. One, the Job-Family Role Strain Scale, concerns general worries about fulfilling both family and work roles. The other, the Job-Family Management Scale, concerns the ease or difficulty people have in managing family activities. (See the scales in Figures 1 and 2. The conceptualization and construction of the scales are discussed in Appendix B.)

Family work, defined as those activities related to home chores and child rearing, is measured by asking people to estimate how much time they spend on chores and doing things with or for their children each day. A weekly average is calculated for each person. (Figure 3 shows the way the questions were asked on the questionnaire; Appendix C explains the conceptualization and history of these measures.)

Family equity is defined as how equally husbands and wives share family work. It is measured by summing the total hours a spouse reports for the family work of self and spouse, and then

FIGURE 1
JOB-FAMILY ROLE STRAIN SCALE

Please indicate by circling the relevant number next to each statement *how often you feel* each of the following:

	Always	Most of the Time	Some of the Time	Rarely	Never	Not applicable
My job keeps me away from my family too much.	1	2	3	4	5	8
I feel I have more to do than I can handle comfortably.	1	2	3	4	5	8
I have a good balance between my job and my family time.	1	2	3	4	5	8
I wish I had more time to do things for the family.	1	2	3	4	5	8
I feel physically drained when I get home from work.	1	2	3	4	5	8
I feel emotionally drained when I get home from work.	1	2	3	4	5	8
I feel I have to rush to get everything done each day.	1	2	3	4	5	8
My time off from work does not match other family members' schedules well.	1	2	3	4	5	8

Please indicate by circling the relevant number next to each statement *how often you feel* each of the following:

	Always	Most of the Time	Some of the Time	Rarely	Never	Not applicable
I feel I don't have enough time for myself.	1	2	3	4	5	8
I worry that other people at work think my family interferes with my job.	1	2	3	4	5	8
I feel more respected than I would if I didn't have a job.	1	2	3	4	5	8
I worry whether I should work less and spend more time with my children.	1	2	3	4	5	8
I am a better parent because I am not with my children all day.	1	2	3	4	5	8
I find enough time for the children.	1	2	3	4	5	8
I worry about how my kids are while I'm working.	1	2	3	4	5	8
I have as much patience with my children as I would like.	1	2	3	4	5	8

Please indicate by circling the relevant number next to each statement *how often you feel* each of the following:

	Always	Most of the Time	Some of the Time	Rarely	Never	Not applicable
I am comfortable with the arrangements for my children while I am working.	1	2	3	4	5	8
Making arrangements for my children while I work involves a lot of effort.	1	2	3	4	5	8
I worry that other people feel I should spend more time with my children.	1	2	3	4	5	8

calculating the percentage done by each spouse. Families with the most equity are those whose sharing is closest to fifty-fifty.

The questionnaire was pre-tested among fifty federal employees in three agencies with characteristics similar to the sample groups. A few modifications in the scale items in syntax and skip patterns were suggested by the pre-testing. Most important, the reliability of the two scales to measure stress was established (see Appendix B).

In its final form, the forty-two-question questionnaire booklet was printed in two versions, one for the flexitime group, which included fourteen additional questions on their experience with the program in their agency (thirteen pages), and the other for the standard time group without the flexitime questions (ten pages). The content included demographic questions to establish the background characteristics of the groups (reported

FIGURE 2
JOB-FAMILY MANAGEMENT SCALE

On days when you are working, how easy or difficult is it for you to arrange your *time* to do each of the following? *Circle the relevant number for each activity.*

	Very Easy	Some-what Easy	Nei-ther Easy nor Diffi-cult	Some-what Diffi-cult	Very Diffi-cult	Not Appli-cable
To avoid the rush hour	1	2	3	4	5	8
To go to work a little later than usual if you need to	1	2	3	4	5	8
To go to health care appointments	1	2	3	4	5	8
To go on errands (e.g., shoe repair, post office, car service)	1	2	3	4	5	8
To go shopping (e.g., groceries, clothes, drug store)	1	2	3	4	5	8
To make telephone calls for appointments or services	1	2	3	4	5	8
To take care of your household chores	1	2	3	4	5	8
To help or visit neighbors or other friends	1	2	3	4	5	8
To participate in community activities	1	2	3	4	5	8
To adjust your work hours to the needs of other family members	1	2	3	4	5	8

On days when you are working, how easy or difficult is it *for you* to do each of the following? *Circle the relevant numbers for each activity.*

	Very Easy	Somewhat Easy	Not Easy or Difficult	Somewhat Difficult	Very Difficult	Not Applicable
To have meals with your family	1	2	3	4	5	8
To spend fun or educational time with your family	1	2	3	4	5	8
To take your children to health care appointments	1	2	3	4	5	8
To take your children to or from a child care setting or school	1	2	3	4	5	8
To go places with your children after school	1	2	3	4	5	8
To go to school events and appointments for your children	1	2	3	4	5	8
To make alternative child care arrangements when necessary (e.g., school snow days)	1	2	3	4	5	8
To be home when your children get home from school	1	2	3	4	5	8
To stay home with a sick child	1	2	3	4	5	8
To make arrangements for children during summer vacations	1	2	3	4	5	8
To have relaxed, pleasant time with your children	1	2	3	4	5	8

FIGURE 3
THE QUESTIONS ASKED TO MEASURE FAMILY WORK

The following group of questions is related to the *home chores* in your family—things like cooking, cleaning, repairs, shopping, yardwork, keeping track of money and bills—plus planning and arranging for all that has to get done.

How many *hours* a day do each of the following persons spend on home chores like those listed above?

	On days when working	On days when not working	Not applicable
You			
Your spouse			
Your children	(on school days)	(on non-school days)	
Others in the household Who?			

The following group of questions are related to the time you spend taking care of or *doing things with your children*—things like feeding, dressing, washing, going places, helping with homework or projects, disciplining, talking, reading, driving them places, etc.

On the average, how much time do you spend on any or all of the above?

On days when you are working? ______ hours per day
On days when you are not working? ______ hours per day

On the average, how much time does your *spouse* spend taking care of and doing things with your children like those listed above under Section C.

On days when he/she is working? ______ hours per day
On days when he/she is not working? ______ hours per day

above), and the measures for the dependent and control variables (discussed below). The cover page of the booklet consisted of a letter to the participants from the Family Impact Seminar that explained the purpose of the study, invited employees to participate, and insured them of the confidentiality of their answers. Work schedules were not mentioned explicitly. (See Appendix H for copies of the questionnaires.)

Data Collection

Based on pre-test familiarity with federal employees, their office buildings, and work environments, a procedure was devised for distributing and collecting the survey questionnaires that aimed to avoid both the impersonality and low return rates of mailed questionnaires, and the large time and cost requirements of personal interviewing, yet include the low-cost advantages of mailing, and the high-response rate advantage of interviewing. This collection procedure included the following steps: introductory letters, distribution, and collection.

Introductory letters. Letters were prepared for the agency heads to send to each employee and supervisor, informing them that the agency was cooperating with the George Washington University in this study; that employees were invited to participate but that their participation was voluntary (in accordance with Privacy Act requirements); and that the study director, or one of her associates, would stop by with the questionnaire in the following week.

Distributors. Six local people were selected to distribute and collect the questionnaires, four women and two men, who were judged as able to work well in the setting on the basis of their maturity (mean age was thirty-eight), appearance, sense of responsibility, and friendly, self-possessed style.[2] In a one-day training session for the distributors, the purposes of the study

[2]Edgar Boling, Carol Doolan, Ellen Mulroney, Delores Royston, Jane Scott, and Randy Swisher.

were explained, the questionnaire content discussed, delivery and pick up of the materials role-played, and answers to possible questions from employees rehearsed. The distributors worked the equivalent of five working days, giving out questionnaires in the first three days and picking them up on the next or a subsequent day.

Return rate. In the flexitime agency, 85 percent of the distributed questionnaires were returned (463 of 542); in the standard time agency 83 percent were returned (341 of 413). The yield confirmed the general hope that this hybrid collection procedure would capture some of the advantages of both mailing and interviewing (i.e., low cost plus high returns). It was an efficient and unobtrusive method that generally created a positive feeling about the study among respondents and encouraged some to engage the distributors in discussions of substantive issues. In addition, the distributors were able to make observations about the work environments. After the data were *cleaned* (eliminating incomplete forms, and so forth), there were a total of 706 questionnaires useable for analysis, that is, 74 percent of those originally distributed. (See Figure 4, which shows the number of questionnaires at each stage of data collection.)

Data Processing

Using standard practice for processing the data, all of the following procedures were followed: developing a code for the data; training four coders who worked full-time for four weeks to code the 804 questionnaires;[3] eliminating the incomplete or inconsistent questionnaires (29); keypunching data for the useable 706 cases (3 cards for each questionnaire); developing a file, cleaning the data, storing the data on tapes, and processing the information on the George Washington University computer.

[3]Roselyn Dixon, Bonnie Oglenski, Gary Peck, and Jane Scott.

FIGURE 4
RETURN RATES FOR EACH AGENCY

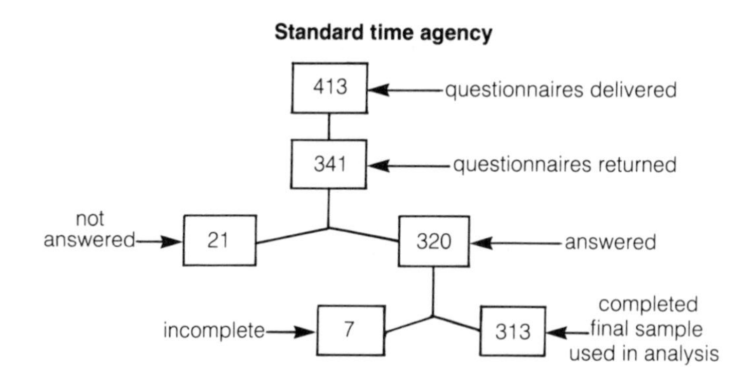

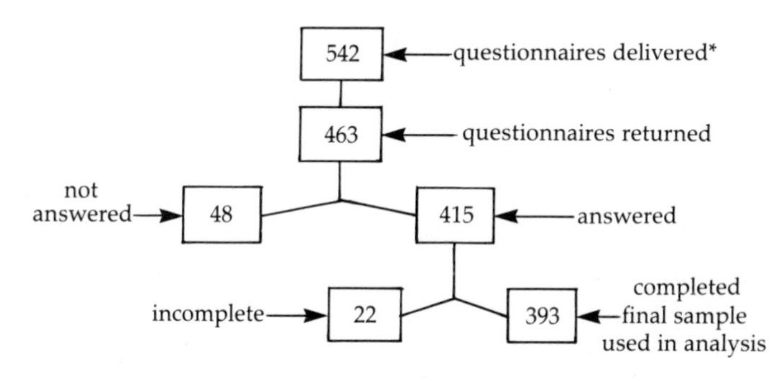

Grand total 706 useable questionnaires were collected from both agencies (74 percent of questionnaires originally distributed).

*Number of questionnaires delivered in the flexitime agency for each type of example:

Original random sample	406
Additional women GS 11 and above	30
Additional parents	106
Total	542

Data Analysis

To test for the hypotheses, the differences between the two agencies were compared according to the dependent variables. The comparisons were intended to respond to three basic questions:

- Do the respondents in the two agencies—taken as a whole—differ on the dependent variables?
- When the groups are divided by family subgroups, what are the effects of schedule on the dependent variables?
- Can the effects of schedule on the dependent variables be better understood if several additional family and work characteristics of respondents are also considered?

To answer the first question, the scores on the dependent variables of all respondents from each agency were compared by a difference of means test (*t* test). To answer the second question, the total sample was divided into husband and wife families (N = 378) and single adult families (N = 276). Each of these was divided into parent and non-parent groups. Within the husband and wife family groups, men and women were considered separately, according to whether both husband and wife worked, or whether only the husband worked. Within the single adult group, single mothers in the two agencies were compared but there were too few single fathers to compare; and single adults without children were compared. Figure 5 represents the analysis plan schematically, by family subgroups, with the relevant dependent variables indicated for each group.

To answer question three, the effect of the independent variable was measured, along with several other control variables, on each family group for which the particular control variables were relevant. These control variables included:

- Total number of hours worked weekly.
- Total number of hours commuted weekly.
- Family life cycle stage.

FIGURE 5
FAMILY SUBGROUPS* AND DEPENDENT VARIABLES FOR EACH GROUP

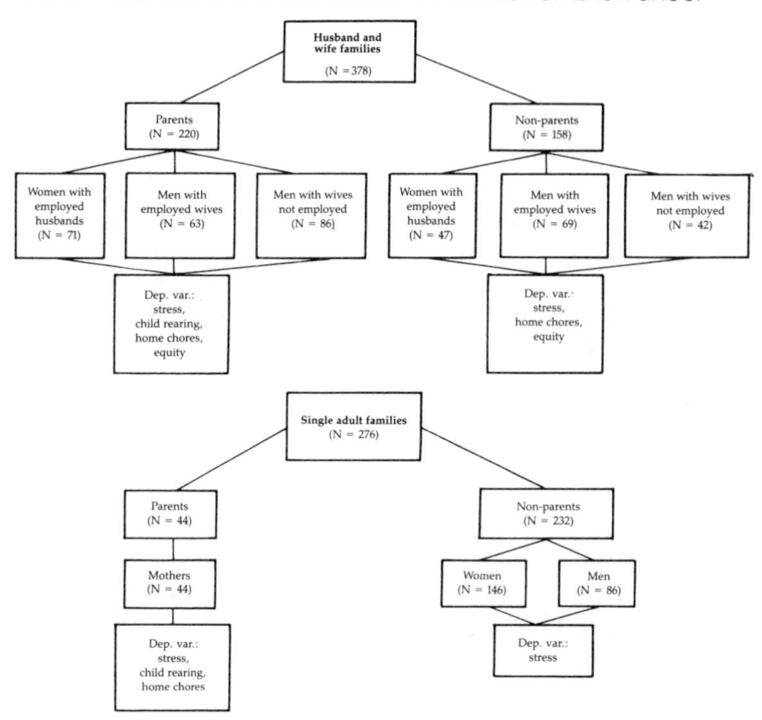

*There were too few women with husbands not employed and too few single fathers to analyze their responses to the questions.

- Occupational level.
- Outside help with family work.

For information on the statistical techniques used for the analysis see Appendix I.

Limitations

The limitations of the study design, which were recognized in advance, are listed below. In addition, several other limitations became apparent in the course of collecting and analyzing the data; they are discussed in the Afterword.

The interrelatedness of work influences. Attempting to isolate work schedules among the many aspects of employment that can affect the family lives of workers is, to some degree, an artificial exercise—especially in a study utilizing an ecological approach to human experience. The schedule may be a relatively unimportant dimension of work influences in particular families, as compared with personality, job satisfaction, sex-role attitudes, job opportunities, family structure, and the like. Moreover, no matter how successfully the sample federal agencies are matched in the study, schedules can never be wholly separated from other factors influencing work and family interconnections, even for research purposes. Thus, whatever effects on families may, or may not, appear to result from different work schedules must be viewed cautiously as only part of highly intercorrelated phenomena.

The process of stress reduction. The study does not aim to explain the process through which schedule may ease the stress involved in being both an employed worker and a family member; yet knowing how stress may be reduced is ultimately an important aspect of understanding why flexible schedules may be helpful to some people. Also, the study does not attempt to investigate alternative ways of alleviating stress related to work and family conflicts, for example, through part-time scheduling, leave policies, and child care facilities.

Information from only one parent. Most contemporary couples research indicates that spouses tend to give somewhat different accounts of objective household activities and issues. For example, in this study husbands or wives may overestimate or underestimate their own or their spouse's family work. Our hope was that the variations would cancel each other out, but the uncertainty about possible direction of error remains.

Personality differences. The study does not attempt assessment of personality differences, although clearly such variations among people will make enormous differences in their use of time and feelings of stress. Again, the hope was that the differences would balance each other out to some degree.

No direct information on children's well-being. Without access to the children and home environments of the federal workers in the study, no direct evaluation can be made of the comparative benefits of the two work schedules on children. But the underlying assumption of the measures created for the study is that working parents who are less stressed and share their family work more equally will rear their children with more satisfaction and more success.

Superficiality of information. A short-answer, self-administered questionnaire cannot elicit in-depth understanding of the complex and varying effects which different work schedules may have on families. At best it will suggest general patterns whose substance and quality will vary among individual families. Yet the survey may provide some indication of the way in which a structural workplace innovation may foster a social change, in this case in the way in which the time and timing of work affects families.

Generalizability of findings. Since the sampling of federal employees was purposive rather than random (for reasons explained earlier in the chapter), the results can be generalized only to workers in the two selected agencies. However, the findings may suggest some family effects of flexitime programs in general, especially for similar samples of federal workers in Washington, D.C., or workers in other urban bureaucracies.

6

The Work Setting

A SENSE of stolidity and inertia pervades the immense and impersonal fifty-year-old headquarters building of the Commerce Department, where thousands of people do endless amounts of routine paper work, week in and week out, year in and year out. Although in some offices and at some levels in the hierarchy work pressures are acutely felt, for the majority of tenured employees in these two agencies studied, there is a homogeneity of experience, and an almost unshakeable sense of job security. The executive branch is also relatively immune from the rising or falling fortunes of both market and political factors in the nation. Change occurs slowly in this large-scale work system; and individual efforts blend imperceptibly into the whole. The 700 people in our survey spend eight hours a day, two hundred fifty days a year, in the environment described and pictured below. Almost all of them are tenured civil servants. This impression of their work setting provides a basis for helping assess what difference a little flexibility in the daily work schedule may make in their lives.

The horizontal order of federal Washington—with its wide streets, green malls, and low, long massive buildings—dictates a slower pace than that of midtown Manhattan, for example, where narrow streets, quick cabs, and fast-rising elevators sustain the hurry-up tempo of the tall, thin skyscrapers. To walk from one first-floor corner of the Commerce Department to its diagonally opposite eighth-floor corner—more than three blocks away—takes fives times as long as a ride in an elevator to the top of the Empire State Building.

Forty-five hundred Commerce Department employees commute an average one-and-a-half hours a day between their suburban homes in Virginia, Maryland, and the District of Columbia and their offices; half ride in carpools of at least four persons because they wish to be eligible to park in the lot across from their building; the rest ride the commuter buses or the clean, quiet, two-year-old subway system (lately becoming crowded at rush hours).

Like many other Washington federal office buildings, the nearest neighbors of the Commerce Department are also massive neoclassic structures. At present the nearest shops and restaurants lie several blocks away, across four- and six-lane avenues. However, two acres of pavement at the north end of the building are presently torn up and surrounded by a construction fence in preparation for a major redevelopment project to enliven the outdoor pedestrian life of the area. By 1981, the automobile-dominated intersection at the tip of the Federal Triangle will become a double plaza—with warm weather restaurants, grassy park areas, and dozens of new trees. On the north side of the plaza, the old Willard Hotel will be luxuriously restored, and a large Marriott Hotel constructed.

When six thousand workers began coming to their jobs in this location in 1932, it was the largest office building in the world. Contemporary observers declared it "a building triumph" of "inspiring beauty" that looked as though it might have been "hewn from the Rock of Gibraltar." It was praised for the visual simplicity of its neoclassical design (columns and central porticos on each facade), and for the engineering complexity of its construction. The architects solved novel structural challenges to build it on the Tiber River bottom, thus making it the first of the major buildings to occupy L'Enfant's "Federal Triangle"—

the long wedge of land, bordered by Pennsylvania and Constitution Avenues, between the White House and the Capitol (A building triumph 1931; Department of Commerce 1975).

What visual impression does the interior of this former showplace among Washington government buildings give? Since the energy crisis in 1973, one has to become accustomed to the dim artificial light before being able to see the tall, dark green pilasters guarding the grand space of the two-story main lobby, or the detailed coffers of its ceiling, or the ornate brass chandeliers hung high overhead. Uniformed guards at the circular desk inside the front door check everyone's purpose in coming—but anyone can enter, just by giving a name and a room destination to the list-keeper at an adjacent desk.

Dominating a handful of unrelated lobby displays—including signs about "America's New Look"—the illuminated "Census Clock" ticks off the numbers of the ever-growing population of the United States. Half-a-dozen new potted palms look healthy in spite of the dark. Part-way down the corridor opposite, a barber shop pole twirls slowly, incongruously, in front of a concession that was opened in the Nixon administration as part of an effort to help minority businesses. In the subterranean cavern beneath the lobby, hundreds of exotic fish stare bleakly in the tanks of the National Aquarium. Adjacent to this quiet water-world, the best cafeteria in any federal building serves 3,000 to 4,000 lunches a day—in a windowed but unadorned three-hundred–foot dining room with courtyards on either side.

Eight miles of corridors—running throughout seven stories—seem everywhere the same: long, wide, empty, and silent. There are a few exceptions: the sixth-floor corridor, with a snack bar and mailroom, has more people and more noise. Office doors are usually closed. Occasionally people cross from one door to another, or round a corner to an elevator or restroom. The endless hallways seem designed for a larger scale and volume of traffic; walking through, one feels like a lone pedestrian

in the Lincoln Tunnel. The rolling robot mail carrier varies the landscape twice a day, blinking and ding-donging as it stops to announce the arrival of mail to a blank office door. The initial complexity of the vast space soon becomes a predictable pattern for a newcomer. But except in the library, and in the computer center, just walking the corridors gives a visitor little sense of what goes on behind the vista of office doors.

Inside the offices of the two agencies we visited, another physical pattern became apparent: secretarial desks are just inside the corridor doors; bosses sit in inner offices. Two to twenty

persons occupy these complexes of little rooms, each set housing a division or bureau within the agency. Larger, higher GS-level executive offices dominate the eighth floor corridor. Occasionally, as in accounting, thirty people work at adjacent desks in a big open space, within sight and sound of one another. Countless telephones and stacks of printed paper exhibit the mechanics of bureaucratic work—everywhere. Many offices are colorful, clean, neat—like glossy pages in office furniture catalogues. Very few reflect the personal interests or family lives of their occupants: few photographs, drawings, posters, or birthday cards are displayed. The predominant mood conveys a sense of lazy order and routine; no hurrying, no crowds, no surprises. Few people seem to smoke or drink coffee at their desks.

What work goes on in this immense, stolid federal office building? Standing on the sidewalk in front of the main entrance, with his or her head thrown back, a passerby can read only a few words of the inscription chiseled under the cornice and running nearly the length of the limestone facade: " . . . within this edifice are agencies to buttress the life of the people, to clarify their problems and coordinate their resources. . . ." This expansive declaration found its echo in a contemporary manufacturing journal, which proclaimed the Commerce Department building to be "a temple of truth dedicated to enlarging the science as well as the practice of business[,] . . . an institution unique in the world to help commerce and industry. . . . It is a great fact finding institution . . . where data from any country or state will pour in to be digested and made useable . . . to foster and promote foreign and domestic commerce" (A building triumph 1931). In contrast, forty-seven years later *The Washington Post* characterized the same governmental edifice as

"an eight acre mountain of stone and cement[,] . . . the equivalent of your Aunt Flo's attic—a jumble of items that no longer fit into a streamlined decor, yet are too valuable to do away with" (Sinclair 1978). Quoting one of the department's own assistant secretaries, the *Post* reporter went on to call it "a backwater of the federal bureaucracy[,] . . . a place where you put things when you didn't know what else to do with them."

Since 1903, when the department was formed, the building has housed and dispersed a variety of government functions and services including aviation and communications regulation, the Department of Labor, the Bureau of Public Roads, the Coast Guard—and more. Today its fourteen major components (some housed elsewhere) include: the Bureau of the Census, the Bureau of Economic Analysis, the Economic Development Administration, the Industry and Trade Administration, the Office of Minority Business Enterprise, the United States Travel Service, the National Oceanic and Atmospheric Administration, the Maritime Administration, the National Telecommunications and Information Administration, the Bureau of Standards, and the Patent Office (Sinclair 1978).

One of the two agencies for our study is among the oldest of these, the other among the youngest in the department. They occupy opposite ends of the building. Both are grant and loan operations, but products of different political and technological eras, designed to serve different needs.

The Economic Development Administration

The creation of the Economic Development Administration (EDA) in 1965 reflects the mid–twentieth-century American concern with how the federal government can assist communities to alleviate unemployment and low family income in their areas. Within the mandate of the Commerce Department to "stimulate business," the existence of EDA represents the contemporary concern to help economically underdeveloped areas of the country develop their potential. To this end, EDA makes loans and grants to local state governments, and to businesses, for various kinds of public works—for example, water treatment facilities, power plants, or industrial parks, which in turn are expected to enhance communities' ability to attract business and create jobs.

Nationwide, EDA employed about 1,050 people in the fall of 1978 (including about 250 short-term employees), 450 of whom are in the Commerce Department headquarters. The EDA professional staff includes mainly economic development specialists, financial analysts, civil engineers, and community planners, as well as lawyers and economists who have specialized knowledge of economic development. Many have worked previously in the public sector at the non-federal level and are sometimes characterized as "socially concerned like those who work for the Peace Corps or HEW."

EDA's fortunes and responsibilities have oscillated with changing administrations. Following a 1973 proposal to phase out the agency, many people left. Currently a pending plan would increase EDA's budget from $708 million to $1.7 billion, largely through new programs in the area of development financing, including $90 million in guaranteed loans. In general, EDA enjoys strong support at all levels of government since it relies on local and state economic development plans when deciding which grants and loans it should approve. Moreover, the Carter administration has made it the major economic development agency by clearly defining its role vis-à-vis other federal grant and loan programs.

In terms of its current volume of work, EDA has an image, within the department, of being understaffed; our distributors also heard complaints of overwork in some offices. One administrator thought it was unlikely that flexitime could make any difference in EDA at the present (if it were started), because program responsibilities keep people longer hours anyway; he has two secretaries covering his office ten hours a day, one coming early and one staying late. This impression was corroborated by the questionnaire data, which indicated that EDA employees work an average of two hours a week longer than Maritime Administration employees.

The Maritime Administration

While the Economic Development Administration distributes grants and loans to projects ranging from sewers to beaches to minority businesses—in every state of the union—the Maritime Administration activities are focused exclusively on the ocean-going American shipping industry. The agency's mandate and $600 million budget are intended to promote a strong, self-reliant merchant marine for the United States. In practice—outside of wartime—this means that MarAd programs essentially provide direct assistance to ship-operating and construction companies through a variety of subsidy, insurance, and tax programs. These programs essentially help maintain a fleet of American flagships—that is, vessels built, operated, and registered by United States citizens—in the face of competition from foreign shipping interests. Of the ten major ship-operating companies in regular common carrier service in deep ocean trades, seven receive regular MarAd subsidies; and most of the twenty-seven ship-building yards in the country—including the twelve to fourteen largest—receive subsidies. Five hundred eighty flag-ships currently operate; eighty-three new vessels have been built since 1970.

In addition to these subsidy activities, MarAd sponsors shipping research and development, supports marketing programs to increase shipper patronage of American lines, provides technical assistance for port development, fosters minority employment and business in the maritime industry, and trains merchant marine officers at the national and state academies. After the agency's reorganization in 1961, its regulatory functions were assigned to a separate regulatory commission.

Historically, the agency has also maintained merchant ships and seamen for defense needs. During World War II, in its earliest incarnation as the War Shipping Administration, the agency trained merchant seamen in a nationwide program that was a mirror image of its military counterpart. During the Vietnam War it mobilized 170 ships (now moth-balled) to transport 95 percent of the war materiel.

To maintain its operations today, MarAd has about fifteen hundred employees nationwide—about eight hundred in the Commerce Department headquarters building and the rest in four regional offices and at the Merchant Marine Academy. About half of the professional MarAd personnel have come to the agency with technical backgrounds in marine engineering or architecture, or from ship-operating positions. (Many have the title of "Captain" from prior shipboard roles.) The other half of MarAd professionals most often have economic or legal backgrounds, and tend to work in the subsidy programs, where they are trained as maritime specialists. The predominantly male composition of the traditional maritime industry is reflected in the fact that MarAd has a smaller proportion of women at high GS grades than EDA. The mean female grade in EDA is 10 whereas in MarAd it is only 6.

The general impression of those walking in and out of dozens of offices to distribute and collect questionnaires was of organizations functioning in a calm and orderly fashion without great pressure, for the most part. Some professionals and clerical people seemed involved, busy, and harried; some seemed idle and restless; the majority seemed to do their jobs in a matter-of-fact, even-paced way. Most were cooperative and pleasant, and willing to take time from reviewing memoranda, taking phone calls, filing papers, typing, and sundry other bureaucratic activities to fill out the questionnaire for this study. Occasionally, people shared reactions to the questions with distributors. Their views varied: some single people felt that the questions were not related to them; a few people thought the study was worthless;

others thought the topic was important; some wanted to discuss the themes, for example, how hard it was to work and have time for one's children.

How Flexitime Was Started in the Maritime Administration

From the beginning of the Carter administration, the top administrators in the Commerce Department—former Secretary Juanita M. Kreps, former Under Secretary Sidney Harmon, and Assistant Secretary Elsa Porter—were identified with efforts to "humanize" the workplace. For example, the Working Life Project, run by Robert and Margaret Duckles in two Commerce Department offices, under contract with the Harvard Project on Work, Technology, and Character, was modeled on the program the Duckleses had conducted for Sidney Harmon in his automobile mirror factory in Bolivar, Tennessee. The Duckleses' Commerce Department program sought to involve civil servants in decision-making processes for improving their work lives. In this climate, in early 1977, Russell Stryker and other MarAd administrators found support for the introduction of flexitime in their agency.

This was ironic because the major impetus for the introduction of flexitime had actually come several years earlier, in the winter of 1973–74, when fuel shortages and traffic congestion stimulated interest in devising commuting alternatives to rush-hour congestion. At that time, the idea of schedule flexibility in a massive bureaucracy was still viewed (by the Civil Service Commission and others) as too risky an undertaking, in terms of both administrative disorder and the potential reduction in productivity.

With more experience in other federal agencies and encouragement from the Commerce leadership, this fear had subsided by 1977 and the MarAd program was approved to begin in September of that year. The reasons for starting the program were still based on an interest in easing commuting problems, but the broader rationale for flexitime—discussed in earlier chapters in terms of increasing workers' sense of control over their lives, raising morale, and so forth—had become part of MarAd's official statements of purpose by this time.[1]

[1]The information in this chapter was obtained from the printed materials cited in the text and from observations, brochures, memoranda, newspaper articles, annual reports, and interviews with the following: Robert Baldesaire, Government Services Administration, assistant manager in the Department of Commerce; Edna Bee, assistant to the director, Office of Personnel, Maritime Administration; Robert and Margaret Duckles, Working Life Project, Department of Commerce; Walter Farr, former chief counsel, Economic Development Administration; Lorin Goodrich, deputy director, Office of Management and Administration, Economic Development Administration; Peter Hannums, employee relations officer, Office of Personnel, Maritime Administration; George Havener, former classification officer, Office of Personnel, Department of Commerce; David Larkin, former acting director, Office of Administrative Services, Department of Commerce; Gerald Lucas, confidential assistant to the assistant secretary for administration, Department of Commerce; Peter McClintock, auditor, Office of the Inspector General, Department of Commerce; Ward Sinclair, *The Washington Post*; Anthony Stadeker, director of administrative services for the Department of Commerce; Russell Stryker, assistant administrator for policy and administration, Maritime Administration; Freda Waddell, acting chief, Property and Buildings Management Division, Office of Administrative Services, Department of Commerce.

7

The Survey Findings

AS discussed in the foregoing chapters, the survey aimed to test the assumption that flexitime helps people achieve a satisfactory balance between their work and family lives. Among the family-related hopes for flexitime were that it would relieve stress related to work-family interfaces; that it would increase the time people spent in "family work"; and that it would help equalize sharing of family responsibilities between men and women.

The experiences of federal workers in a flexitime agency (MarAd) were compared with those in a standard time agency (EDA) on the three issues. When there was a significant difference of means between the two agencies on any of the three dependent variables (on stress scores, on hours in family work, or on percentage of time spent in family work), we then conducted a multivariate analysis to see if the differences remained after other possible influences (like family life cycle stage, occupational level, and total hours worked and commuted) were taken into account. The results reported in this chapter are the same for both the difference of means tests and multivariate analyses, unless otherwise noted. (See Chapter 5 and Appendix I for information on the multivariate methodology used; data tables for all the measures are presented in Appendix J, Tables 12–23.)

Family Stress

At first glance at the data, the expectation that people on flexitime would have less work-family-related stress seems well founded: the mean stress level of the whole group of flexitime employees is significantly lower than that of the whole group of standard time workers. Again, when families with children are looked at separately the results are still promising: parents on flexitime still have significantly less stress (see Tables 12 and 13).

Moreover, when the respondents are divided by sex, the results are still encouraging. Specifically, it was anticipated that women would be especially helped by flexitime in terms of balancing their job and family lives. The whole sample of women on flexitime did indeed report less stress on both stress scales (see Table 14). Furthermore, all the employed married women with employed husbands reported experiencing less stress if they were on flexitime (see Tables 12 and 13).

Thus, the first look at the findings suggests that the hopes for flexitime are borne out: flexitime women and parents report significantly less stress than their standard time counterparts.

For men alone, however, the results are more mixed. When all the flexitime men are compared with all the standard time men, the flexitime group has significantly less stress on the Family Management Scale but not on the Role Strain Scale (see Table 14). But when married men in both agencies are compared, the flexitime group has no reduction in stress, whether their wives are employed or not (see Tables 12 and 13).

Moreover, when the parents are divided by sex, the advantages of flexitime in reducing stress dwindle further. For mothers, they disappear altogether. The mothers on flexitime—married and single—do not have less stress than those on standard time. The fathers are helped slightly more: they have less stress on the Family Management Scale, but not on the Role Strain Scale (see Figure 6, and Tables 12 and 13).

When the parent groups are looked at even more closely, in

FIGURE 6

EFFECTS OF FLEXITIME ON STRESS IN VARIOUS FAMILY STRUCTURES (ON THE ROLE STRAIN AND FAMILY MANAGEMENT SCALES)*

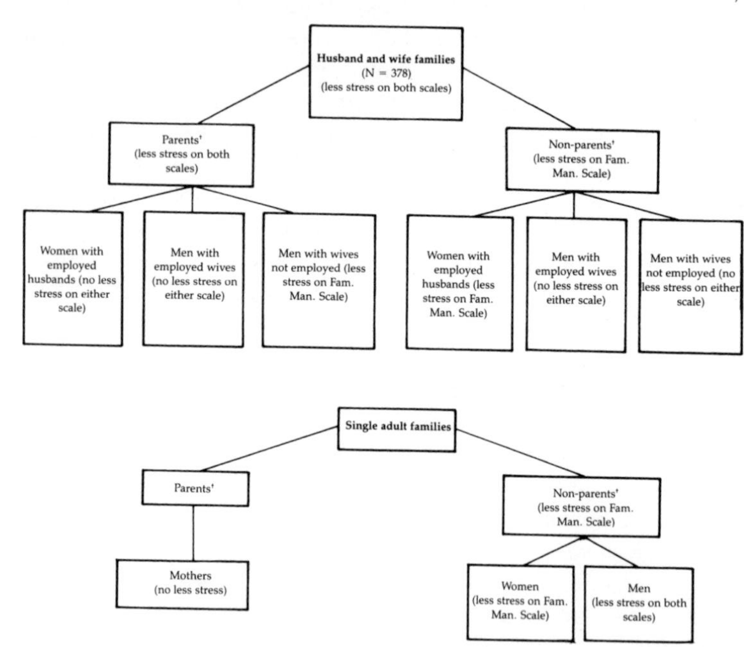

*For data tables, see Appendix J.
'There were too few women with unemployed husbands and too few single fathers to analyze the results for these groups.

terms of whether the respondent has an employed spouse or not, flexitime seems to ease the stress of only one category of parent, namely, fathers whose wives are not employed. This group has significantly less stress on the Family Management Scale but not the Role Strain Scale. Both mothers and fathers who have employed spouses, as well as single parents (mothers only in our sample), have no less stress if they are on flexitime (see Figure 6, and Tables 12 and 13).

For husband and wife families without children, flexitime is more helpful in reducing stress for women. On both of the stress scales these wives on flexitime have significantly less stress than standard time wives without children. But the non-fathers on flexitime, with employed or non-employed wives, have no less stress than standard time non-fathers (see Figure 6, and Tables 12 and 13).

Of all the groups without children, flexitime was most helpful in reducing stress for single adults. The men had significantly less stress on both scales and the women less stress on the Family Management Scale (see Figure 6, and Tables 12 and 13).

For the stress measures in this study, the most telling finding is that the flexitime workers who have less stress than their standard time counterparts are individuals without primary responsibility for children, namely, married women without children, fathers whose wives are not employed, and single adults without children. The primary group whose family lives were expected to benefit from more flexible work schedules, namely, employed mothers, did not report less stress than those mothers on standard schedules.

However, for two of the groups with less stress on the Family Management Scale, flexitime was much more important in explaining the lower stress scores than the other variables considered in the multivariate analysis. For flexitime fathers whose wives were not employed, schedule was more important in explaining the lower stress scores than family life cycle stage, total hours worked, and occupational level. Similarly, for married women without children, schedule explained more of the variance than the total hours worked weekly and occupational level. (To reduce the quantity of data in this volume, tables for the multivariate analyses are not included in Appendix J, but they may be obtained from the authors. Similarly, additional comparison of means tables are also available.)

The fact that childless women (i.e., without children under 18 or in the home) on flexitime have less stress than their standard time counterparts, whereas mothers on flexitime do not have less stress than those on standard time, may indicate that the modest schedule flexibility in the agency examined is sufficient to help the first group but insufficient to help the second (in the ways measured in the study). In short, it seems likely that the complexities of combining work and family life for employed mothers probably far exceed those of employed women without children—and a modest flexitime option apparently does not help much in dealing with the related stress.

The lower stress scores on the Family Management Scale for married women without children may also be related to subjective and objective aspects of stress. The non-mothers on flexitime (married and single) report less stress in relation to the logistical or "management" aspects of their lives; but they do not have less stress than their standard time counterparts on the subjective or "how it feels" dimensions. In other words, flexitime may make it objectively easier for them to coordinate and accomplish a range of personal and family-related activities, but they may still feel strains in the process of accomplishing these functions.

But the fact that this difference does not also occur for men without children (nor for fathers whose wives are employed) also raises questions about another aspect of the objectivity and subjectivity. In terms of respondents' interpretation of the question "how easy or difficult is it for you to arrange your time to do . . . ?" it is unclear whether the scale taps the objective difficulty people think they have in accomplishing the particular activity or task, or whether it measures their subjective reaction to simply having those responsibilities.

For example, a man who has not typically engaged in most of the activities on the list as part of his family life may assume that it would be very difficult for him to do them; thus, his report that it would be difficult to do them may reflect the fact that it has seldom occurred to him that he should or could do them (e.g., grocery shopping or household chores). Women, on the other hand, may answer more objectively, in the sense that they more typically assume responsibility for family activities, and therefore can more realistically assess how easy or difficult it is to do these things on days when they are working.

One of the most surprising results of the study—in the sense that attention was least focused on this group by the congressional testimony and other literature—was the fact that single men on flexitime have significantly less stress on both the scales than standard time men. Single women also had less stress on the Family Management Scale, suggesting that flexitime provides significant benefits for single adults in managing the logistical aspects of their lives—more benefits, relatively, than married people or single mothers gain from flexitime in this respect. (Allan R. Cohen also reported, in his 1978 study of alternate work schedules in the John Hancock Company, that single people seemed to benefit the most from flexitime.) But the fact that single women on flexitime feel as much role strain as those on standard time suggests that schedule flexibility does not ease the more psychological complications of balancing work and personal/family life.

Family Work

Our hypotheses for the family work measures were more exploratory than predictive. That is, for men, the implication from the congressional hearings, and elsewhere, was that those on flexitime would increase their time on family work; for women it was not clear whether their actual time on family work would go up or down. When the total sample of men is considered together in a simple difference of means test without controls, the flexitime group does average significantly more weekly hours on home chores than the standard time group (16 versus 14 hours at $p \leq .05$); the whole group of flexitime women also averages significantly more weekly hours on home chores than their standard time counterparts (26 versus 23 hours at $p \leq .05$). However, in all other groupings, the most consistent finding on this variable was that most workers on flexitime did not spend significantly more time on family work than those on standard time. This was true for the total group of respondents, as well as for the total group of parents and non-parents on both home chores and child rearing (see Figure 7, and Tables 5 and 6).

In the family subgroups, none of the flexitime parents spent more time in child rearing than the standard time parents. And only the single mothers on flexitime spent significantly more weekly time on home chores than their standard time counterparts (29 versus 21 hours per week; see Figure 7, and Tables 15 and 16).

Even in the multivariate analysis, the flexitime single mothers spent eight hours a week more on chores than standard time mothers—and schedule explained twice as much of the difference as family life cycle stage. Thus, it seems likely that schedule flexibility facilitated increased time on chores (see Table 17). However, the fact that the stress level for these single mothers was no less than that of the standard time group suggests that the additional time on chores did not provide much relief from the overall sense of pressure felt by these mothers in managing both jobs and family responsibilities.

Although the difference in hours spent on home chores between the flexitime and standard time married women (mothers and non-mothers) is much smaller than the differences between the two groups of single mothers—and not statistically significant—a similar trend may be apparent. The fact that the flexitime married women spend slightly more time on chores than the standard-time married women (one-and-one-half and two hours per week, respectively, for mothers and non-

FIGURE 7
EFFECTS OF FLEXITIME ON TIME SPENT IN FAMILY WORK IN VARIOUS FAMILY STRUCTURES*
(ON HOME CHORES AND CHILD REARING)

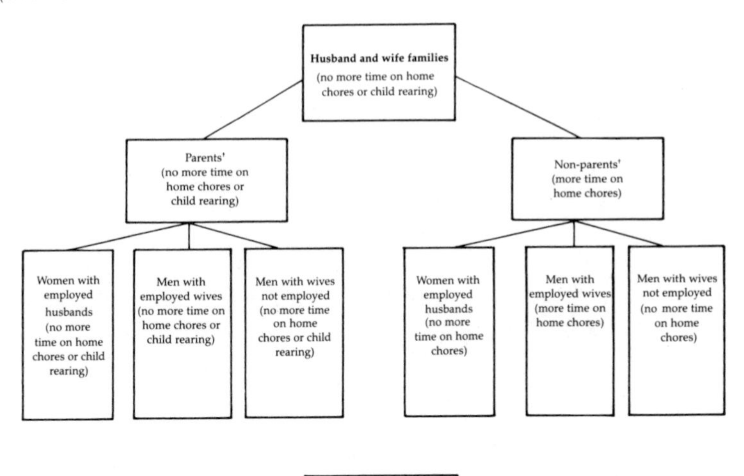

*See data tables in Appendix J.
†There were too few women with unemployed husbands and too few single fathers to analyze the results for these groups.

mothers) may be viewed as a negative effect of flexitime when viewed in conjunction with the equity question. That is, with respect to the male-female role questions addressed in the study (see Chapters 2, 3, and 4, and Appendix E), one effect of flexitime appears to be that flexitime mothers wind up not only doing more total hours of chores per week, but the married ones also do a larger share of the chores vis-à-vis their husbands than the standard time group. Although the married non-mothers on flexitime do also have less stress, the married mothers have no less stress than their standard time counterparts, as well as spending more time on chores.

For men and home chores in this study, another trend is worth highlighting, apart from the non-effect of flexitime. In contrast to earlier evidence on the family work of married men (e.g., Dizard 1968; Nye 1976), the fathers whose wives are employed—on both kinds of schedules—report spending more weekly hours on home chores than the fathers whose wives are not employed. The standard time men with employed wives spend eighteen hours a week on chores while those with non-working wives spend fourteen hours a week; the flexitime men with employed wives spend seventeen hours a week on chores; and those with non-employed wives spent sixteen hours a week.

Pleck (1979) reported a similar finding in his analysis of the 1977 Michigan Quality of Employment Survey, specifically that employed men with employed wives do 1.8 hours per week more in housework than employed husbands with non-employed wives ("the first finding of non-trivial increments in husbands' family work associated with wives' employment in a study assessing family work absolutely in terms of time in a large representative sample"). In our study, however, the men without children whose wives are employed do not spend more time on home chores than those with non-employed wives.

This study also adds further information about the disparities between employed mothers and employed non-mothers, in terms of stress and time on chores. The mothers spend about seven hours a week more on chores than the non-mothers. This is true regardless of schedule (see Table 17). And the mothers on flexitime have no less stress than those on standard time.

Family Equity

The third way in which the standard time and flexitime workers were compared in the study was in terms of sharing family work, that is, to see if husbands and wives divided family work more equally in their families when one spouse had a flexitime option. Like the expectations for family work, our hypotheses for this variable were also exploratory.

As is predictable from the fact that the flexitime men did not spend significantly more time than standard time men on family work, the schedule flexibility does not appear to encourage men to share home chores or child rearing more equally with their wives. On this variable there were no significant differences between the total samples of flexitime and standard time workers. Nor were there any significant differences when the groups were divided by sex, by parental status, by marital status, and by whether the spouse was employed (see Figure 8, and Tables 17 and 18). For example, the men with employed wives reported doing 40 percent of the home chores in their families, irrespective of schedule, while those men whose wives were not employed reported doing about 25 percent in their families, irrespective of schedule.

Thus, while the differences in time spent on home chores for men on different schedules are not significant, the differences are very great, depending on their wives' employment status. In terms of the central questions for this study, the independent variable (schedule) is a much less powerful influence on the dependent variable (increasing equity in family work) than a control variable, namely, spouse employment status.

We explored the implications of these family work sharing

FIGURE 8

EFFECTS OF FLEXITIME ON EQUITY
IN FAMILY WORK FOR HUSBAND AND WIFE FAMILIES

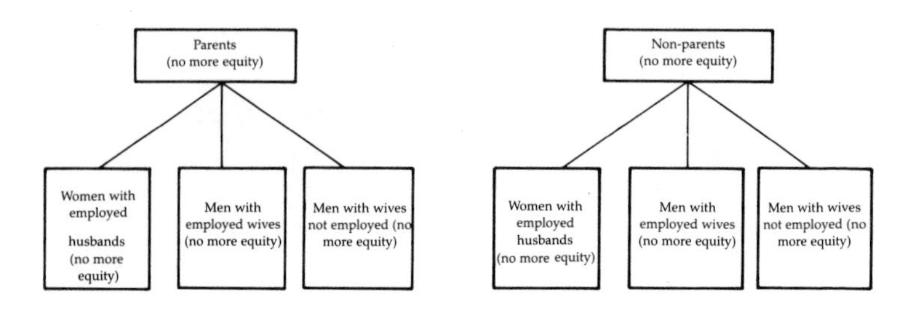

patterns a little further through two supplementary questions, asking, first, who took the main responsibility for family work in each family and, second, whether one spouse wanted the other to do more. While 50 to 60 percent of the men in both agencies said the responsibility was equally shared with their spouses, less than 40 percent of the women said the sharing was equal. Moreover, more than 40 percent of the women, in both agencies, wanted their husbands to spend more time on home chores, while only about 15 percent of the men wanted their wives to increase their time on chores.

Although the particular men and women reporting were not married to one another, and without considering issues like different standards in housekeeping, variations in efficiency, and so forth, the trends are still striking. Three issues seem important. First, even if the men exaggerate what they did (which may be true if the women's reports of men's hours spent on family work are accurate), the employed men still contribute far fewer hours proportionately to family work than their employed wives. Second, these women work and commute just about the same number of hours per week as the men. (See Tables 20 and 21; we have assumed the figures would be comparable for the actual employed wives of the men in the sample.) Third, only about a third of the women want their husbands to spend more time on home chores (see Table 22).

In contrast to these modest hopes, or preferences, of the employed women regarding having their spouses share chores, they appear to have much stronger feelings about having the men involved in child rearing—again regardless of schedule. Close to 65 percent of the women on both flexitime and standard time would like their husbands to spend more time with their children (in contrast to only 40 percent wanting husbands to spend more time on chores). Only about 15 percent of the men on both schedules, on the other hand, want their wives to spend more time with the children (and about 12 percent want their wives to spend less time with the children; see Table 23).

In general the findings on family equity, or sex role division of labor in family work, are consistent with those of the major time budget studies (e.g., Szalai 1972; Walker and Woods 1976) with the exception of the trend discussed above, namely, that men on both kinds of schedules whose wives are employed spend more hours on chores than those with non-employed wives; thus, the family work is shared more equally in these two-earner families (generally 60:40 versus 75:25 in families with non-employed wives—according to the husbands' reports; see Table 17).

But for this study the important finding is that schedule flexibility has no significant influence on the different sharing patterns; rather it is the fact of the wife's employment that seems to account for the greatest difference. Since this comparison was not made directly in this study we cannot be sure of the hunch, but it is a topic worth further exploration by other researchers. Parental time with children—and attitudes of mothers and fathers on this issue—may be one of the most important areas for future study, in light of the trends towards full employment for both parents of most children.

In terms of the conceptual frameworks for this study, an interesting connection appears between the ecological and family systems theories (see Chapter 4 and Appendix B) in the following sense. In ecological terms, it is when an individual (a wife) from a family system (the microsystem) establishes a functional link with the work system (the mesosystem) that the division of labor in the family changes. The policy of the work system examined in our study (flexitime) did not significantly influence the shift in the division of family work. In this kind of structural analysis, however, the work policies and practices leading to the greater employment of women (along with other factors) appear to be major forces leading to changes in the sharing of family work.

In family systems theory, the domestic work roles within the family seem to shift as a result of changes in the behavior of one member of the system; that is, when the wife is employed, the husband does more family work than he does in families where the wife is not employed.

Characteristics and Perceptions of the Flexitime Group

In addition to measuring the differences between the two agencies on the three major dependent variables, the responses of the flexitime group alone were also compared in two ways. First, those who changed their schedules when the flexitime option became available in MarAd were compared with those who kept the regular 8:30 to 5:00 schedule in order to see if they differed in any significant ways in background characteristics or on work-family-related issues. Second, the responses of all of the MarAd group were analyzed with respect to their perceptions of the flexitime program in their agency.

Fifty-seven percent of the flexitime agency respondents in the sample changed their schedules when offered the flexitime option (223 changed, and 163 did not). Also, the agency's own one-year evaluation of its flexitime program found that many additional employees intended to change schedules when their personal circumstances permitted (e.g., carpools and other family members' schedules).

Fifty-seven percent of these "users" also vary their starting times occasionally, or more often; and even among those who did not change their daily schedules, 55 percent vary their starting times occasionally or more often. All the offices in the agency participate in the program; the lowest participation rate in any office is 50 percent, and in some offices everyone participates.

The two groups differed significantly on only two background characteristics: first, a larger percentage of the flexitime group had regular commitments (outside their MarAd jobs) of more than five hours per week, including part-time jobs, volunteer activities, and educational activities, in that order; and second, people who live with at least one other person use flexitime significantly more than those who live alone.

On other background characteristics, there were only a few trends of interest. A slightly larger percentage of males than females changed their schedules (60 percent versus 54 percent), which is contrary to the general expectations in the hearings (but without marital status and family life cycle stage characteristics of the groups this difference is not particularly meaningful in terms of the focus on families). People over fifty years of age used flexitime least; but in all other age groups the user and non-user groups were of similar size. With respect to our interest in families with children, twice as many users as non-users have children under age six; this difference is almost statistically significant ($p \leq .07$). However, the average number of children did not differ between the groups. A slightly smaller percentage of blacks than whites changed their schedules.

The predictions discussed in Chapter 4 about use of flexitime in relation to occupational level and long working hours were confirmed. People who work more than forty hours a week use flexitime less; those with advanced degrees and at GS grades 15 and 16 use it less than those at all lower GS grades. These findings confirm the expectation, discussed in Chapter 4. On the other hand, those at the next lower professional and administrative GS levels (13 and 14) work only forty hours a week, on the average, and more often than not changed their schedules with flexitime. Users and non-users did not differ in terms of general job satisifaction; but those who changed their schedules are significantly more satisfied with their work schedule than those who did not change.

Even though the background and work characteristics of the users and non-users differed little, as indicated above, the groups differed significantly with regard to the ease they reported in doing certain things. Most importantly, in terms of the focus of this study, those who changed their schedules under flexitime were twice as likely to say that "it was easy to adjust their work hours to the needs of other family members" (item on the Family Management Scale, r = .15). Similarly, among their reasons for using flexitime, "having more time with the family" was most important (r = .18), along with avoiding traffic and being able to better schedule personal activities. The users also reported significantly more ease than the non-users in avoiding rush hour and in not getting to work late. According to their questionnaires, very few people used flexitime at lunch; but the agency's own evaluation reported that employees frequently cited the mid-day flexitime band as a beneficial feature of the program because it eliminated "hassles" about long lunch periods.

Despite their positive responses to flexitime for family reasons, as indicated above, the users reported just as much interference as the non-users on the job-family interference question. Joseph Pleck has suggested that this finding may be explained by a self-selection factor, that is, those who choose to change their schedules may well be the group with the most job-family conflicts in the first place; therefore, even if their interference level is reduced, it may still be as high as for non-users.

In response to the question about what prevented people from using flexitime as much as they would like, "the attitudes of supervisors" was the reason on which there was the most significant difference between the user and non-user groups. The agency's own one-year evaluation of the flexitime program blames supervisor problems with flexitime on supervisors themselves: "strong and weak supervisors retain the same character traits and behavioral patterns regardless of the work hours arrangements" (Wells n.d., p. 7). This finding is similar to reports in the flexitime literature (e.g., Hilgert and Hundley 1975), and suggests an area for further investigation with respect to the role of implementation in the success of flexitime programs. The "need to meet with or coordinate with others" was the second major difference between users and non-users in terms of limitations on their use of flexitime.

Eighty-two percent of all the MarAd respondents said that flexitime was successful or very successful; and this percentage was even higher among the users (87 percent versus 75 percent among non-users). When asked what additional schedule flexibility they would like, the most popular choice among MarAd respondents (191 selections) was the four-day week with ten-hour days. The *banking and borrowing system* was the next most popular (140); this system, for which experiments were authorized in the legislation passed in September 1978, allows people to average eighty hours of work every two weeks, while varying their daily totals. Fifty flexitime workers also wanted longer vacations. Only twenty-four expressed preference for more part-time opportunities, and only twenty-one wanted four-day weeks with eight-hour days and less pay.

These reactions to flexitime echo those of many private and public sector evaluations of flexitime programs. Above all, the MarAd employees like their flexitime program; only 4 percent of the respondents think it is unsuccessful. To further understand the effects of the schedule flexibility for various types of families, however, more must be known about the differences between male and female users, in terms of family earning structure, stress, and time spent on family work. However, given the size of the user sample (223), the numbers in the subgroups would be too small to make statistically meaningful comparisons.

To gain further insight on how and why flexible work schedules might be helpful to people, we asked the following open-ended question at the end of the questionnaire (no. 58): "Some people have found that more flexible work schedules have significantly affected their lives outside of work. If your personal and/or family life has been affected by greater flexibility in your work schedule, please describe these changes."

Ninety-four of the 393 MarAd respondents (almost one-quarter) chose to tell us about such changes. Seventy-nine of these wrote that flexitime had been helpful, and offered some examples. About half of these were parents (N = 37). Proportionately more parents answered this question than were in the whole sample; the majority of this group were fathers whose wives were not employed.

These fathers (N = 16) emphasized the following advantages of flexitime: having more family time in the evening and being able to have meals (breakfast and dinner) with the family, especially the children; reducing commuting time (which ranged from ten to sixty minutes for this group); reducing strain and tension; having more time for recreational activities, both individual and family-oriented; and using less leave time. Two of these fathers wrote the following specific comments. A GS 13 non-supervisor with a three-year-old praised the MarAd program, saying it offered "more individual decision-making in a bureaucratic environment, . . . less time commuting, . . . more relaxation, . . . more time with family." Another GS 13 supervisor with twin two-year-olds wrote that flexitime "allows scheduling of unforeseen activities and to be *different*."

Respondents in the other parent subgroups cited the same major reasons for flexitime's advantages in helping with their family lives. Of the forty-two parent respondents, 31 percent (N = 13) also took advantage of the quiet time at work, especially in the early morning.

Of the four major reasons parents (N = 42) specified for liking flexitime, the percentages of parents who mentioned each reason are as follows:

Family/children—50% (N = 21)
Commuting—50% (N = 21)
Use of less leave—50% (N = 21)
Recreation—38% (N = 16)

The parents' comments included the following. A married woman (first marriage) with children ages five and eight, who is a grade 9 non-supervisor wrote: "It seems to enable me to accomplish more—doing dinner, shopping, etc. . . . It also enables my oldest son to participate in more after school activities . . . flexitime is the best thing the government ever initiated." Likewise, a father with four children between the ages of seventeen and thirty-one who is a GS 13 non-supervisor and whose wife is employed wrote: "Absenteeism has been sharply reduced. . . . It has restored more control of my personal and working life to my hands. I now get credit for coming in early[,] . . . the tension of commuting has been greatly reduced[,] . . . I have more family hours[, and] . . . my family finds me more relaxed." And a grade 7 married woman with a seventeen-month-old baby wrote: "I now have time to grocery shop, wash clothes, and spend time reading with my daughter. . . . I find myself tiring much earlier, but after my daughter is bedded down."

A remarried father with children ages ten and twelve who is a GS 12 non-supervisor saw the value of flexitime as follows: "It is basically successful if a person isn't a creature of habit and sees the potential of being able to control his/her work schedule. . . . It makes all the difference in the world if you can get on the parkway by 7:00 A.M. . . . By leaving work early, I can sometimes pick up my kids and bring them to the museums and art galleries, a learning as well as enjoyable experience." And finally, a grade 12 father with children ages nine and fourteen whose spouse is employed cited the following advantages:

"People enjoy the flexibility . . . although it is abused sometimes . . . can arrange doctor's appointments and auto repairs . . . can participate in more after-hour activities."

Among non-parents, the largest number of people who answered the question were single females (N= 20, 54 percent of non-parents answering). A larger percentage of women answered this question than were in the total sample, while proportionately fewer men without children answered this question than were in the total flexitime sample.

Respondents in the non-parent subgroups cited five major reasons for flexitime's advantages in their lives. The numbers and percentages of non-parents who mentioned each reason are as follows:

Use of less leave time—68% (N = 25)
Recreation—63% (N = 23)
Commuting—57% (N = 21)
Quiet time at work—49% (N = 18)
Education—46% (N= 17)

Another fourteen people chose to answer the question in order to explain why they thought flexitime was not helpful (this group made up 15 percent of those who answered the question). Of the eight who were parents, all were male; of the non-parents in the group three were men and four were women.

The comments of these eight fathers primarily explained that the flexibility in the MarAd program, in and of itself, was too minor to make noticeable differences in family life. However, one father said that even with only a little flexibility he saved leave time for when he had appointments at his sons' school; another said that his office director had always been flexible anyway (and that this supervisor would like even more authority to give his supervisees more schedule latitude). Thus, this group of fathers definitely likes flexitime. The importance of their comments is largely that they would like more flexibility.

The comments of the non-parents who answered this question by saying that flexitime did not help much also tended to be very favorable towards the concept, but mentioned several particular problems. For example, one man said that carpool and parking problems prevented his taking as much advantage of flexitime as he would like; another woman said her supervisor was still too stringent; another felt that the theory was right but the implementation was abused in her office.

Thus, of the 25 percent of MarAd respondents who answered this question, half of the parents specified the advantages in terms of family life, as well as reduced commuting time and use of leave time. For non-parents these second two reasons were also the most important advantages of flexitime. In general, people valued flexitime highly for family and other reasons.[1]

[1]The data from the questionnaire includes various kinds of additional information that are available for further analysis. For example, we asked several questions about children: how much time they spend watching television, what kinds of activities parents engage in with them, and how much time the children spend on home chores. The answers to these questions could be considered and related to the major dependent variables, stress and time in family work, and other items.

Similarly, we could look at the correlations between particular items on the stress scales and factors like the family life cycle stage, employment patterns, and time spent in and sharing of family work. We could compare younger and older workers for differences in behavior and attitudes. For example, are younger cohorts of men sharing the responsibility for family life more equally than older men? What difference does the wife's employment status make? Do the ages, sexes, and numbers of children play any role in the division of family work between mothers and fathers? What effects do job satisfaction and job absorption have on the ways family work is divided?

The data set also includes interesting information on household composition and other kinds of family functions. For example, from the answers to the question, "Who lives with you?" we can develop interesting portraits by age, race, and sex of various kinds of households among the seven hundred respondents. One hundred and nine of those who filled out the questionnaire reported that they provided special care for family members or others "due to things like illness, a handicap, or old age." What are the relative amounts and kinds of stress and time on family work for these people?

Finally, one-third of the respondents in the sample are unmarried, and thus we have data on the family relationships, jobs, attitudes, activities, and stress

A Summary of the Findings

The most prominent finding from the survey is that the kinds of workers who were expected to be most helped by flexitime schedules do not seem to benefit much by the modest flexitime program examined in this study—by the measures of family life on which they were compared. On the other hand, by more general measures than the particular dependent variables chosen for this study, flexitime appears to have benefits for family life. In MarAd's own evaluation, the largest majority of respondents said they liked flexitime because they could more easily schedule personal activities before or after work. On the questionnaire for our survey, one-third of the MarAd respondents reported they used flexitime in order to have more time with their families. On another question, half the respondents said that an important reason for using flexitime was to have more time with their families. Also, the flexitime respondents reported significantly less job-family interference than standard time employees. Thus, in terms of employees' own direct perceptions and reactions, they like flexitime, both for work and family reasons, and it seems to reduce work-family conflicts. Moreover, flexitime is being used in many places—not just talked about by reformers. In the case of the Maritime Administration, 91 percent of those who participated in the agency's own flexitime evaluation wanted the program to become permanent; just as many supervisors in MarAd changed their schedule as not; and a mere 4 percent of the MarAd respondents in the survey thought their flexitime program was unsuccessful.

Yet with the indirect measures used in the study the benefits of flexitime for family life are less apparent. Why is this so? The

---

for an unusually large sample of single people. This group includes subsamples of single mothers (N = 44) and divorced persons (N = 84). In the case of the "remarried" persons (N = 67) in particular, very little research has been done on this growing population. What are their relative stress levels, sharing patterns, and attitudes towards men and women's roles in work and family?

survey research design explicitly sought to look beyond the general impressions of flexitime popularity to learn in what ways the policy might help families. It attempted to look at flexible schedules both more conceptually—to consider one aspect of a work system in terms of its effect on several aspects of a family system—and more operationally, by assessing family stress, family work hours, and how equally husbands and wives share family work. These efforts both complicated and muted the more impressionistic expectations for the benefits of flexitime for families.

The results of this application of family impact analysis suggest several important substantive conclusions about the topic, and raise several methodological questions. Plainly, the magnitude of the logistical, energy, and time demands on families with two employed parents, or a single parent, cannot be dramatically altered by minor changes in daily work schedules. Ironically, the reverse may also be true. That is, parents with young children may be precluded from varying their schedules—even when they have a flexitime option—because the logistics of their lives are so fixed. For example, the schedules of the babysitter, child care center, school, or other parent may dictate when they can go to and from work.

In this connection, therefore, the relatively minor leeway for choosing one's working time in the existing MarAd flexitime program (allowing no reduction in the eight-hour day and only half-hour deviations from pre-selected starting and leaving times) probably could not make a major difference in the stress and behavior patterns (i.e., family work) of families like that described in the vignette in Chapter 3. But even with the minor schedule changes permissible in this modest flexitime program, which did not make measurable differences between the two agencies on the variables we considered, the vast majority of the MarAd flexitime workers (96 percent) think their program is successful.

The disparities between the direct and indirect ways of evaluating the impact of flexitime on families can probably be explained in terms of a continuum from simple to complex, in three senses. First, in terms of the inquiry itself, pushing into the spectrum of positive attitudes towards flexitime leads inevitably to points of specificity at which there are diminishing returns from the advantages of this work schedule option. It cannot resolve all work-family stresses for all kinds of workers. Second, people can have positive attitudes towards the idea of choice in the scheduling of their work while still recognizing the limitations of the modest version of flexitime examined in this study. Third, the results suggest that the simpler the family circumstances of the respondents, the more relative impact a little schedule flexibility seems to have in helping them balance jobs and family life. Thus, single adult families without children appeared to have been helped more—in terms of reducing stress—than parental families with two employed parents and young children. Or, in other words, a small degree of flexibility helps a lot with little problems (i.e., the logistics of single adult families); but it helps only a little with big problems (i.e., the logistics of families with children and employed parents).

Both the stress and time-use findings from this study probably raise more research and policy questions than they answer about the relationships between work and family life at this point in American history. Certainly they force us to ask more about the work and family roles for men and women: What combinations are best for both adults and children? How much can or should work policies influence the choices?

8

The Follow-up Interviews

Methodology

EARLY in the process of planning the survey on the effects of flexitime on families, several members of the Family Impact Seminar suggested that we supplement the short-answer aggregate data by means of in-person interviews with a small number of respondents who filled out the written, self-administered questionnaires. Their hope was that through interviews with respondents and their families we could learn more about the general effects of work on family life and thereby better understand the relative importance of work scheduling among other work influences. In open-ended conversations we could seek explanations for the survey findings by pursuing the interconnections between answers to different survey questions by hearing more than one member of a family discuss their circumstances. The resulting information would be less generalizable than the survey data, but would suggest more insightful and complex explanations of the interplay between work and family responsibilities.

Format

Following this suggestion, our original proposal for this second phase of the study included plans for both interviews and observations with respondents and their families. However, after reviewing the recent literature on the difficult methodology of participant observation we became convinced that elaborate and lengthy design and data collection processes would be required to meet acceptable research standards for this technique.[1] Neither time nor funding for the work schedules study permitted a full-blown effort of this kind.

Meanwhile, we heard of an alternative methodology that seemed even better suited to our needs. The Project on Human Sexual Development, funded by the Rockefeller Foundation and based in Cambridge, Massachusetts, had conducted a study in Cleveland on how parents teach their children about sexuality (see Roberts, Kline, and Gagnon 1978). They used interviews to follow up a large survey.

For their interviews, the Cleveland Study authors had experimented with several kinds of parent interviews: with one couple in their home, with two couples together in an interview room, with a group of mothers and fathers in a meeting room. In their written materials and in telephone conversations, the authors reported that among the three formats, the group discussions were by far the most fruitful.[2] The group discussion worked best primarily because the participants found each other's comments mutually stimulating and therefore the discussion generated an abundance of relevant information in a short space of time.

[1]The literature on interviews and participant observation that was reviewed included: Jacobs 1970; J. M. Johnson 1975; Becker et al. 1968; Sudman and Bradburn 1974; Sanders 1974; Golden 1976; Adams and Preiss 1960; Dexter 1970; McCall and Simmons 1969; Denzin 1978; Myers 1977; J. M. Johnson 1977; Sykes 1978; Wiseman 1974; Weiss 1968; Rist 1975; Bailyn 1977; Weiss and Rein 1970. Also included were the following journals: *Anthropology and Education Quarterly; Social Science Research; Quality and Quantity; Measurement and Evaluation in Guidance; Educational and Psychological Measurement.*

[2]In July 1979, we talked by telephone with the project director, Elizabeth Roberts; with the research director, Robert Kline; and with Judith Simpson, executive director of the Cleveland Program for Sexual Learning. We reviewed the mimeographed study materials, including the technical manual and appendices (Roberts, Kline, and Gagnon 1978).

In our review of the research literature we did not find any reports on using group discussions to validate survey data, or to obtain additional information from individuals who had participated in a survey. But we were able to review videotapes of the Cleveland sessions and the discussion guides for the meetings. On the basis of these very helpful materials and our conversations with the authors, we decided to conduct our own interviews with groups of survey respondents.

Purposes

Our purposes for the interviews evolved through several phases, with the following three objectives sustained throughout:

1. To learn more about how people feel their work lives both benefit and conflict with their family lives; and about the relative importance of their work schedules among the ways in which they feel stressed or supported in managing their family responsibilities.
2. To identify additional policies, programs, and services that might ease the work-family conflicts they feel.
3. To evaluate how well our written questionnaire tapped the ways in which the scheduling of work, per se, affects people's roles as family members.

For the first objective, we constructed a variety of possible approaches. For example, we considered asking each participant to describe an aspect of their providing, consuming, socializing, and coordinating behaviors (functions drawn from the Family Impact Seminar framework for analysis, see Chapter 9). In terms of consuming, for instance, we considered asking, "If it seems useful to talk to a child's teacher, who makes the decision, who calls for the appointment, who goes, who acts on the outcome of the discussion?" The purpose of this question, and similar ones, was to gather specific information about how husbands' and wives' work roles might affect such aspects of family functioning.

For the second objective, following discussion of these and similar questions about child care, home chores, and consuming, we planned to ask couples to suggest work policies that would reduce stress or difficulties they encountered in performing their family roles. For the third objective we planned to review each questionnaire item in terms of its applicability to the subject.

In final form, these approaches were simplified into two major questions with the expectation that participants would mention enough specific instances and themes in their own experiences to serve our purposes without our naming examples of activities. These two main questions for the discussions were: What are the benefits and difficulties work creates for your family life? What helps balance your work and family life?

As part of these two general questions we were interested in learning how parents in two-earner families coordinate and divide responsibilities, what the difficult issues are, what goes well, and what helps or causes problems. In these connections, we wanted to know if they could think of ways in which employers, or policies, could better structure work to facilitate family functioning. Because of time limitations we never did evaluate the questionnaire as systematically as we originally hoped, that is by asking the groups to comment on each item.

Interview Organization

We obtained permission to use the videotape facilities of the Family Therapy Institute in Bethesda, Maryland, to conduct the interviews. The Institute is the family therapy training center of Jay Haley and Cloe Madanes, and is located in this Washington, D.C., suburb in a small house with a kitchen and comfortable living room–like accommodations. To fit into the Institute's schedules, we conducted two group interviews on successive evenings in the summer of 1979, ten months after the survey data were collected. Following the most successful of the Cleveland Study formats, we invited the participants to come after work, for drinks, dinner, and discussion (from 5:30 until 9:30 P.M.). All of the families were reimbursed $25.00 to cover expenses incurred in their participation.

We developed a detailed plan for the evening, including explanations about the purposes of the research, about confidentiality, about rules of the discussion, and a procedure for introducing participants, and for asking a sequence of discussion questions. We pre-tested the discussion plan in a group session with eight members of the Family Impact Seminar staff.[3] We revised the plan afterward with the help of their suggestions.

Selection of Participants

On the basis of the survey results, we considered a number of different ways to select respondents for the group interviews in relation to particular variables. Initially, we planned two groups that would be chosen on the basis of the following independent, dependent, and control variables from the survey: schedule (flexitime), stress level, sex and earning status, family life cycle stage, and number of children.

The first group was to consist of mothers on flexitime from two-earner families and their husbands. Each of these mothers was to have more than one child, and at least one child under twelve years of age; and she was to have a high stress score on our scales. Our sample included twenty women in this category. The second group was to consist of single mothers (separated, divorced, or never married) with the same characteristics as the mothers from two-earner families. Our sample had seventeen women in this category. We intended to begin with these two groups both because the congressional testimony on flexitime gave so much emphasis to its predicted effects for mothers, and

[3] Elizabeth Bode, Darlene Craddock, Mary Eng, Ruth Hubbell, Sidney Johnson, Laurie Keyawa, Theodora Ooms, and John Sheldon.

because our survey data suggested that mothers did not seem to be helped much by the modest Maritime Administration flexitime program on our measures of stress, family work, and equity.

We also thought, however, that several other groups of survey respondents would provide equally interesting interview data on work-family relationships. For example, we wanted to explore why the fathers on flexitime whose wives were not employed showed more differences from their standard time counterparts than most of the other comparison groups. In another instance, because one-third of our sample consisted of single people, and because little is known about the relationships between single adults and their extended families, we thought it would be valuable to interview respondents in this group.

While continuing to focus on parents with relatively young children and high stress scores, we finally decided that the opportunity to make comparisons across several variables was the most important criterion for forming the discussion groups. First, we concluded that it was important to have people on both kinds of schedules in the same groups in order to benefit from the direct comparisons they would make with one another about their schedules. For similar reasons, we also wanted people at low and high job grades in each group. Third, we decided that each group should include both men and women from the two agencies so that we could hear about the differences and similarities for males and females within and between the respective agencies.

We finally decided that we would interview families in which both husbands and wives were employed, because this family type, along with single mothers, seemed an important contemporary family structure, increasing in number, for which balancing work and family life was complicated. Knowing more about their needs and solutions to two-earner parenting problems might provide the most useful new data on work and the family at this point. However, by trying to represent so many characteristics in each group we would not be able to make many comparisons between people with similar characteristics.

On the basis of these criteria, we constructed a list of about one hundred respondents from both agencies. We designed two groups of ten that would be mixed in the ways described above, that is, fathers and mothers from both agencies at high and low GS grades, blacks and whites, all with young children and high stress scores. Responses to our letter of invitation (Figure 9) were very favorable. Although most people (97 percent) said they would like to come, only about half were able to accept the invitations, and some had to withdraw later for health reasons or other conflicts. We continued to invite from our lists, trying to keep the groups balanced as planned. We sent follow-up reminder notes and called each family before the discussion. The meeting finally included one group of nine and one group of four. (Two families in the second group were unable to come at the last minute.)

The eventual composition of the groups was skewed from our survey sample in terms of race and job grade level. Among the thirteen participants there were no blacks (although several accepted initially); and only one non-professional grade employee (a secretary). Thus, the final interview sample was biased in terms of socioeconomic and education level, as reflected for the most part in the job grade. In turn, the spouses of these respondents also tended to be high-level professionals. All were members of intact families; only one came without a spouse. Five of the participants were from the flexitime agency (three men and two women); and two from the standard time agency (one man and one woman). The six spouses worked in the private sector.

Given the composition of the groups—that is, predominantly on the upper end of the job grade and socioeconomic spectrum—the interview data do not add much to our information about the ways in which the Maritime Administration's flexitime program affects less educated and less prosperous families.

FIGURE 9
COPY OF INVITATION SENT FOR INTERVIEWS

INSTITUTE FOR
EDUCATIONAL
LEADERSHIP

July 30, 1979

THE
GEORGE
WASHINGTON
UNIVERSITY

Suite 310
1001 Connecticut Avenue, N.W.
Washington, D.C. 20036

Samuel Halperin
Director
(202) 676-5900

Education Policy Fellowship Program
(202) 676-5925

Educational Staff Seminar
(202) 676-5949

The Associates Program
(202) 676-5935

"Options in Education" over National Public Radio
(202) 785-6462

Education of the Handicapped Policy Project
(202) 676-5910

Family Impact Seminar
(202) 296-5330

Washington Policy Seminar
(202) 676-5940

Fellowships in Educational Journalism
(202) 676-5901

Expanding Opportunities in Educational Research
(202) 676-3970

National Policy Fellows in Education of the Handicapped
(202) 676-5910

Dear :

Last October you very kindly filled out the questionnaire for our study on work and family relationships. As you may remember, we mentioned in the cover letter that we would contact a few people again at a later date to talk with us further about these issues in their own lives. I am writing now to invite you and your spouse to participate in such a meeting.

You are invited for dinner and a discussion on Monday, August 6 from 5:30 to 9:30 p.m. at 4602 North Park Avenue, Chevy Chase, Maryland (see enclosed map). Three other Commerce Department employees and their spouses will be in the group. All of you are parents of young children. We are interested in learning more from you about how employed parents with young children handle their work and family responsibilities: What are the benefits? What are the difficulties? With your permission, we will videotape the discussion for our use in reviewing the issues which emerge from your comments. We will pay you $25 for any babysitting and transportation expenses you may incur. To insure your privacy and the confidentiality of your participation, only the research staff will see the tapes and your name will not appear in any reports of the study.

I will call you within the next few days to see whether you and your spouse can join us on August 6th. Both the discussion and the dinner (sandwiches) will be informal and we expect that you will find the occasion interesting and enjoyable. Thanks very much in advance for your help. I look forward to talking with you further.

Yours sincerely,

Halcy Bohen
Director, Work Schedules Study
Family Impact Seminar

HB/me

Enclosed: Map showing location of discussion
New York Times article describing the Family Impact Seminar

The very fact of the sample bias, that is, that the families who were willing and able to come to the interviews were in the upper brackets of the job and income levels, reflects the limitation of our interview strategy and setting: the realities of the lives of the others may have made it more difficult for them to find the energy and interest necessary to make arrangements—child care and otherwise—for an evening's visit to a white middle-class suburb for a research project. Ideally, we should have done a second set of interviews to reach this group of people in a more effective way. Their stories of work-family issues undoubtedly would have differed from those which follow here.

Interview Findings

Three hours of videotapes were made for each of the group discussions. Each of these was transcribed and edited. The data were analyzed by cutting up the trancripts and organizing the quotations by couple in order to create a portrait of the work-family experience of each participating family. The information in each portrait was organized according to themes—for example, how work affected family life, how mothers' and fathers' roles in work and family differed, and how education, parents, and values affect their current situations and choices. We telephoned a few participants in the days following the dinner meetings to check information and to further explore some themes raised in the group sessions.

The following account of the interviews is organized by portraits of each couple; the analysis of their comments is interspersed with descriptions and quotations from each participant. Their names and other identifying information have been changed. Overall conclusions are drawn about the effects of different kinds of work on different kinds of individuals and families, highlighting the differences between men and women across work and educational levels. The importance of work schedules is assessed in relation to other work factors which were emphasized in the discussions.

The Garretts

As a GS 14 economist, Louise Garrett feels her job in the federal Maritime Administration is less pressured than her husband's work in private law practice; and her hours are shorter and less carefully monitored. At thirty-seven years of age, she is a small, trim woman whose determined demeanor becomes more relaxed and smiling in the course of an evening's conversation. She feels that MarAd's flexitime was very important to her last fall when she had "no stable child care situation" for her ten- and eight-year-olds. But since she received a promotion, Louise sees her situation as follows:

> I could not work flexitime now because I have a supervisory position. I couldn't leave before five. You have to be there for the people who work for you. Or maybe it's for the people you work for. I'm not sure which. . . . But if you accept a job at a certain level you are expected to behave differently. No one ever told me. I wouldn't even ask to go on flexitime. . . . It's not policy, it's attitudes.

Her husband's law firm keeps track of its attorneys' work time in six-minute segments. Clients are billed accordingly. Each lawyer is expected to bring in a certain amount of business each year. In this environment, with an Ivy League B.A. and LL.B., thirty-eight-year-old Ken Garrett says:

> I have a fairly stressful job. Lots of eggs in the air at the same time. It's great for not getting bored. But when I get home my mind stays on my work. It's difficult to unwind with the kids and get into their world. Only occasionally can I relax with them. . . . I have a great deal of flexibility about when I can come and go but in terms of overall balance, I think I have too much on the work side . . . and I also think there isn't a damn thing I can do about it.

Speaking with authority and quick gestures, Ken tries to convey how and why his job requires so much time and energy: "Unless you're there most of the time you can't get the work done. Actually, it's an internalized work ethic. You feel you're not doing a proper job and rationally you know you're not going to get it all done unless you're there."

Although they both felt that their jobs compete with the family for time and energy, the issues were not posed in exactly the same ways for Louise and Ken.

> *Ken*: From a male perspective it is not either/or family or work. It's both. As a practical matter it is not open to sacrifice the career or the family, . . . at least you're damn crazy if you do. . . . In that direction lies suicide, divorce, drug addiction.
>
> *Louise*: He means if you ignore your family.

Ken is certain that to be without a family life would, for him, be disastrous: a lonely and depressing life. Yet he sees no acceptable way to reduce the time and absorption demands of his career.

After a couple of hours discussing these issues with six other employed parents, Ken came up with an explanation for why he had "trouble staying near the right balance" between work and family life—and why the balance usually tips in favor of work:

> In the law firm you have wide latitude about when you put the hours in but not about how many there are. . . . The work makes clear, objective calls on you and the penalties if you don't meet them are clear and obvious.
>
> There are a lot of satisfactions to be garnered from married life and a lot to make you happy about it . . . but the demands, requests, pleas that your family gives you are not so clear and obvious. And the penalties aren't quite so immediate. That tends to tilt the balance towards work. . . . But . . . you [also] feel bad if

> you fail to discharge family responsibilities well. It's a question of balance. In my case, it is impossible for me to strike a balance that I'm satisfied with. . . . I don't know how you fight that except by staying conscious of it and that doesn't really solve the question of how you stay conscious of it, . . . which is why I said at the beginning that there is no solution.

Each couple also addressed the work-family question in terms of how they thought their children were affected by their parental roles in work and family. In all cases the discussants thought the children's expectations of their mothers and fathers were different; and concurrently that the children were learning that employment has a different meaning and a different importance in the lives of men and women.

Ken described his impressions as follows:

> I've taken my kids to work. They see what I'm doing. But it's hard to explain to a ten-year-old what a tax lawyer does. . . . When they see me doing some kind of work they can understand, it tends to be around the house—and I do a crappy job. . . . I don't know what my kids are getting from me in terms of an attitude towards work. I'm not sure they're getting anything very favorable from me. They see me working at home occasionally, at night or on weekends. Whether that's a positive experience . . . I kind of doubt it.
>
> My unavailability doesn't bother the kids. Louise's unavailability does bother them.
>
> *Louise*: That's because they always accept the fact that he's at work. I was home [not employed] until the oldest started kindergarten, and until the other started nursery school. My son suggested that if I wanted to work, I get a job at the "7-11" a couple of hours a day.
>
> It's hard to explain what an economist does!
>
> I'm not sure my children are getting anything from me in terms of an attitude towards work . . . They

> view my job as an adjunct, an irritant to my main responsibility for them. How many times have they called you at work, Ken?
>
> *Ken*: Maybe once in the last year.
>
> *Louise*: They call me three to four times a day—to settle fights, anything. . . . They would never question whether he should be doing his job.
>
> I get guilt complexes. I change my schedule all the time. I worked three days a week last fall. I spent a morning a week in each of their classes last fall. They want me to do it again. They're insatiable.
>
> Women have it harder. There may be cases where there is an equal split of family responsibility. I don't know of any.

Not only did the Garretts start with different work and family roles—she staying home to care for the children while he worked long hours in law practice—but even now that Louise is employed the differences are sustained in terms of what she feels she ought to be doing as a mother. Her children, in turn, expect her to be more active than their father in their day-to-day activities. The corollary of this expectation, in the Garretts' view, is that their children think their father's work is more important than their mother's—and not interruptible for their needs.

The availability of flexible work schedules for both Garretts is far less important in determining this pattern of who plays what work and family roles than three other factors: first, that their long job hours should not be regularly interrupted by routine children's needs; second, that this work ethic is more binding for men than for women; and third, that for women, the family demands are more salient than for men.

In group discussions the seven families divided roughly into two groups on the question of whether balancing work and family responsibilities was difficult or not. On one end of the spectrum, maintaining a satisfactory balance was a constant "juggling act"; at the other end, the participants felt little conflict between work and family roles, and viewed them as "separate worlds."

The characteristics that distinguished the two groups had less to do with their work schedules than with several other factors. The couples who felt they had to struggle daily to sustain an acceptable balance between work and family had both of the following characteristics: two children (pre-school or pre-teen), two demanding and absorbing jobs. Those who perceived few conflicts between work and family had the following characteristics: no pre-school or pre-teen children, and/or the job of one or both spouses was not very demanding or absorbing.

Two other factors also distinguished the two groups of families. Those who felt more work-family conflicts were in higher socioeconomic brackets (as were their parents) and had more internalized drive for achievement and recognition in highly competitive work settings. Overarching all of the above were the major differences both in men's and women's levels and kinds of responsibilities at home and at work, and in the ways work and family conflicts were perceived by mothers and by fathers.

In the first group of families, where balancing work and family life is complicated, flexible work schedules are welcome—indeed taken for granted. But they play a very minor role in the ways work-family conflicts are perceived or managed. Five of the people in this group have flexible schedules: two work for the government and three in the private sector. The one participant on a standard schedule is a federal employee from the survey control group.

The overriding complication of their lives is the magnitude of the combined expectations—internal and external—in their work and family lives. They attempt to sustain challenging work lives and "quality" parenting. Both goals are time-consuming. Responsibilities on neither side are easily curtailed.

None of these couples has chosen one of the strategies suggested by the Rapoports for balancing family and work life successfully, namely, that people either have children while they are young and before launching careers, or that they establish themselves in their careers before having children (Rapoport et al. 1977). Instead of choosing between these alternatives, the couples described here are doing both simultaneously—rearing children and establishing careers. Moreover, they would argue that the options suggested by the Rapoports are unrealistic, at least in some careers; for example, that establishing a career and then reducing involvement in it is not possible for people who wish to advance in that field because only sustained, high involvement and long hours of work will maintain an established career.

The Telfords

In an Eastern urban attorney's uniform—dark suit, muted tie, and white shirt—Warren Telford looks serious and anxious, almost diffident when he appears in a group. He relaxes as he talks, candidly, articulately, carefully, with long, thoughtful sentences explaining the complexities and nuances of his feelings about how his work affects his family life. He smiles cautiously as he speaks, and broadly in response to others.

The Telfords' family and work circumstances are similar to the Garretts'. Like Ken Garrett, Warren Telford is an Ivy League alumnus who works in a private law firm. His wife also works for the government. The Telfords' son and daughter, now thirteen and eleven, are in private school in the District of Columbia; the Garretts' children attend school in the Virginia suburbs, where the family moved for the higher quality of the public schools. The Garretts now have a forty-five-minute commute by car, each way; the Telfords commute thirty minutes each way by bus.

The Telfords' self-portraits of their stress in a dual career family with two pre-teen children are very similar to the family picture sketched by the Garretts. For the Telfords, as for the Garretts, the work and family roles differ for the father and the mother—despite the fact that this husband and wife are both professionally trained in the same profession. What accounts for the differences?

Susan Telford, age thirty-six, returned to her Midwestern hometown after graduating from an Eastern women's college. After several frustrating years in a routine job in an insurance firm, she began law school, got married, had babies, and finished her law degree. Thus, she got her legal training while rearing two young children. Susan says she was "juggling" parenting and career demands simultaneously from the beginning of her professional life. Warren, on the other hand, was already a practicing lawyer before he became a father. Both Telfords feel tugging between their work and family responsibilities, but the form is different for each lawyer-parent.

> *Warren*: I think Ken's point about the balancing process is a critical one. Day in and day out you are being torn by your job and your duties as a parent. I feel that to the extent that there are shortcomings in what I personally do—I have made a bad choice by accepting a certain caseload, or travel commitment—it's just bad planning, just bad. Given the choice to make the right decision, I made the bad decision. You constantly have to make choices.
>
> A law firm atmosphere is highly competitive. Lawyers are ranked against one another constantly. That follows you through your career. There's no such thing as a secure position in a law firm. There are expected numbers of billable hours at the end of the year.

Warren seems torn between feeling that he personally fails when he does not do as much parenting as he feels he ought to; and feeling—in a resigned way—that the structure and ethic of his work world are immutable. The implication is that if he reduces the time and absorption he gives to his work, he will fail in the eyes of his professional peers, and therefore in his own view. On our stress scales, both Warren and Ken had high scores; but the reasons for their stress are not fundamentally explained by their work schedules, which are flexible; their work ethic, sustained by their work environment, is far more important.

Susan, on the other hand, "compromised" her work commitment from the start—and still does. She is a GS 12 in the Economic Development Administration—the standard schedule agency. As with the Garretts, the differences between these two lawyers on this central issue are explained, in large part, by sex-role expectations for men and women; these internalized expectations heavily influence their personal choices and their feelings about their work and family roles.

> *Susan*: It is related to where you start in some respects. I had stress from the beginning with two pre-school kids when I was in law school, worse than now. . . . I would mentally turn off the switch when I got in the car to drive home from law school. I've never had the luxury of throwing myself into something for twenty-four hours that way.
>
> I think it is a strength now that I have survived, as opposed to a young person eager to please, who would give the job everything. I have always had to have a balance in contrast to a twenty-two-year-old whose parents are paying for law school so she can work seven days a week. I never had that.
>
> On Wednesday, for years, I stayed in bed, to recuperate and to be alone. I didn't *have* the time, I just took it. I was exhausted. I stayed in bed alone—with the law books.
>
> Now it's like rolling off a log compared with that. It [young kids] precluded me from having that stressful a job. . . . At the time I was looking for a job I wouldn't have been considered for a stressful one (with two young kids). They excluded me. And I don't have as stressful a job as Warren and Ken now.

Thus, for Susan, her parenting has been a brake on developing a certain kind of law career—for example, becoming a partner in a prestigious firm like her spouse—apart from the question of whether the law firm would have wanted her as a partner. And for Warren, the acceptance of the work ethic and demands of that kind of career leave him puzzling regularly over what the content and process of his parenting should be. Speaking quietly and seriously to the other parents in the circle, he offered the following philosophy of his parenting:

> I have, as everyone does, concepts of how the family should work. My rankings, beyond the physical, financial, and health factors, are that parents must be a presence, be available. Children must be able to pick up the phone and get to me and my wife if we're not there. . . . Time with the children must be directed to some object. Time with them can be wasted, thrown away. Since you have a limited amount of time with kids, what are you trying to accomplish? You want to make it valuable.
>
> I think it is important that kids participate in their own activities for their own sakes, not just for the approval of Mommy or Daddy.
>
> What kind of child are you trying to create with the little bit of time you have? What imprint do you want to leave on the child? . . . Parenting can have some impact on children but it won't ultimately influence what that child will become. Over-involvement is self-delusion. . . . I can justify my sort of lack of involvement in my children day to day in the sense that I am quite convinced that from the day they emerge from the womb they are distinct personalities and they can be helped along and given some values and given some discipline but in the final analysis. . . .
>
> I don't know if I am harming my children or not. I can't tell. I have a kind of guilt feeling. I don't know

where it comes from. Maybe from the way I was raised. I've given up. I try not to let it bother me. I have no way of measuring. . . . My dad was away from home a great deal—travelled one hundred thousand miles per year as a national executive. I've reflected on that, not wanting it to recur for my children. But reflecting on the reflection, I'm not sure, maybe I'm rationalizing, but I'm not sure what difference the quantity of parenting makes. I've seen a lot of bad parenting, fathers and mothers. I feel less guilty now. . . . My problem is that I don't know what quantity of parenting is right—whether being certain places with a child, and being available and so forth, what difference it really makes.

Work can be an escape from the family. There is always one more thing to do as a lawyer—you can cop out, be a workaholic, be obsessive.

Among the themes in Warren's comments, perhaps the most prominent is uncertainty. The quandary is not primarily the magnitude of the multiple demands in his life, but more fundamentally what values should guide his daily and yearly decisions about how to balance his job and family roles. Yet for all his disclaimers, by his wife's testimony Warren has been deeply involved in parenting.

Susan: I am not one who wrote [in the questionnaire] that I wanted my husband to do more. Warren spends high quality time with the kids and always has. When they were babies he was on the floor eyeball to eyeball with them and has come up that way. He relates to the children extremely well. Better than I do. . . . When I was in law school, [chores were done by] whoever was around. I'd fix dinner. He did the dishes and put the kids to bed, did laundry and yard work. I studied on the weekend.

> *Warren:* I changed my career to avoid being away six weeks at a time.
>
> We try to plan separate time with each child because together at these ages they fight. We do one-to-one for a few hours on the weekend. But it doesn't satisfy their cravings built up over a week for time with us. But it evens things out over some, eases the tension. If I don't figure out a way to divide them, nothing is accomplished.

When the career trade-offs, or costs of the "female" versus the "male" role are posed for Susan, she struggles to explain what the differences mean in her life:

> *Do you see yourself ever having a career as demanding as Warren's?*
>
> *Susan:* I wonder about that. I . . . there . . . are . . . It isn't my whole life and neither are the children or home. There are a lot of things I would like to try or take up again. I'm not compulsive about anything. In my work, I'm a generalist and I like that.
>
> *So what's the difference between you and Warren? You're both lawyers, you both have the kids . . .*
>
> *Susan:* Nobody made me feel I had to be a success. It's the only difference I can think of. Nobody gave that message to me. In fact, I got flack when Warren's energy was diverted to the family by my being in school. I was undercutting his chances for success. We moved to Washington because I was looking for my first law job. It was not the best place for him alone. This is not seen as the proper approach. The pressure came from Warren's parents. . . . My parents were also upset that I was not staying home with their darling grandchildren. I was totally unprepared for that. I'm still getting over it. I had never disappointed them before. It never occurred to me that they would be willing to fi-

> nance a legal education for me while I was single and then be adamantly opposed to my finishing.
>
> I had to accept from the start that not everything is worth doing well, either at home or at work . . . [in order] to make all of this work. You have to decide what will be done very well, and what half-assed, and what in-between. If I see my daughter crying in her room, I have to attend to that; when the music teacher says every girl in the chorus has to have a long skirt made by her mother, I say no. At work, I choose to get some things out fast, and to do others really well. I'm not trying to please anybody except me.

While Warren struggles with uncertainties about what kind of parenting is enough, Susan explains the countervailing parental and cultural signals that have shaped her female compromises in balancing work and a family. Anchors for her values no longer come from her parents. Nor do they now come from her workplace. She looks back resentfully on the relentless law school schedule that could not acknowledge her parental demands; and she wishes her government job allowed more flexibility. Susan does not seek more sharing of family work from her husband, who does share. She balances work and family needs by countless compromises to meet enough of the demands from each "system" to satisfy herself. Hers is the "hectic and fragile lifestyle" Haldi describes (see Chapters 1 and 4). A flexible work schedule would be appreciated; but for Susan Telford, the dilemmas are more deeply rooted in confusing contemporary values about what men and women should be doing in work and family.

The Bensons

Thirty-year-old Carolyn Benson looks too young and frail to be a university professor—a scientist—and the mother of two young children. But when she speaks, firmly, articulately, and confidently she seems instantly mature and competent. When three-month-old Andrew interrupted the group discussion too loudly—with coos or cries—Carolyn, or her husband, Steve, matter-of-factly let him suck one of their fingers; or lifted him easily from his infant seat and took him out of the room for a while.

Steve's curly brown hair, and slim face and frame, make him, too, seem a young thirty-year-old father. Although Carolyn left the room more often to care for Andrew, Steve, with a blue baby-sling around his neck, seemed as comfortable as Carolyn with parenting. In fact, he cared full time for their daughter during six months of her infancy, before he found a government job as a budget analyst.

For this young couple, the youngest in the group, with the youngest children, the uncertainties about how much of their time and talent should go to their family and how much to their work is as persistent and puzzling as for the Garretts and Telfords. Both Bensons are committed to their professional careers and they worry whether they spend enough time with their children—and each other.

At the time of the survey, Steve worked in the flexitime agency in the study as a GS 12; at the time of the discussion he had recently shifted to a budget analyst job in HEW. He felt about his flexitime the way the male attorneys felt about theirs: that it was only marginally useful to men like him who were committed to advancement in a chosen career. Despite flexitime, Steve says, "Up the chain there is pressure to stay late. So much work gets done between 4:30 and 6. It's as if they say, 'Okay, we diddled around all day, now decisions have to get made.' . . . It's not policy, it's attitudes. Older people in high GS levels, 15–16, are less likely to accept changes."

But Steve also carefully acknowledges that he cannot easily buck the work ethic and expectations of his job, because it matters to him how his colleagues view his performance. He will be judged, in part, by his willingness to put in long work hours and these judgments are important to his self-esteem: "I can't accept not being successful, whatever damage it does to my family. Somewhere in there I won't take an extra hour off. I will stay late. My family is more important, but when it comes down to how much time you'll put in, it is easier for men to say 'job first.'"

Like the other couples, Steve and Carolyn feel there are differences in the ways the family and work pulls affect men and women. Steve says, "I accepted that I would go to work. Carolyn had to decide. Males just accept: you go to work. Women worry about whether they are abandoning their children, whether they are being evil. A man just goes."

Both Steve and Carolyn also think their own outlook on work and child rearing emerged from their parents' expectations—and that some of these expectations, especially for daughters, were mixed, resulting in some confusion for their adult children:

> *Carolyn:* I was an only child. My father [a chemical engineer] was determined I should have a career. Since I have had children he has pulled back in terms of pressure on me to succeed. I sense personal confusion in his mind about which he would like to see me doing: succeeding in work or being home with my children.
>
> *Steve:* The important thing is parents' attitudes. If they encourage a daughter to have a career they can't just turn around.
>
> *Carolyn:* We know a couple where she is the larger earner. In our case, Steve didn't work for six months and took care of our daughter. But he didn't want to be a househusband.

Like the Telfords and Garretts, the Bensons also express uncertainty about just what kind of and how much parenting is appropriate, right, and good for their own fast-growing babies. Steve worries that he ought to be "there" for his kids while they are available to him: "Pre-school years are the only time when children are without a peer group. It's a very important time for child-parent interaction. When children are older they're interacting with siblings, or TV. Maybe they reject you, too. Therefore, it is very important to spend time with them as babies."

Commenting on flexitime in relation to her parenting, Carolyn offered the following picture of the persisting dilemmas:

> I essentially have flexitime. Some days I leave my daughter and do not see her until the next day—except when she's asleep. It's nice to have a long day off but when I work long days I resent missing my child for two whole days. For the "couple part" of family life it wasn't bad. But for her—or for me, as mother—it was hard to get back into relaxing with her.

Reflecting the complexity of demands on young parents with serious careers, Steve suggested another reason why flexitime alone cannot alleviate all the competing time demands: "Even if flexitime allows parents to devote time to the children you may be losing the parent to the parent. The only time we can talk is in the car. A two-year-old will not permit conversation between two adults."

For all of these professional couples one striking perplexity underlines their work-family concerns; namely, the disparity between the clarity of what is required to be successful on the job side, and the uncertainty about what is required to be successful on the family side. Given the unequal dictates from the two arenas, flexible scheduling as an option on the work side does little to tip the balance in favor of time with the family.

All six of the parents seemed buffeted by changing and conflicting advice and values. In addition to the attitudes absorbed from their own childhood experiences and from their parents' views, they hear conflicting evidence from the "experts," too. These highly educated parents have impressions of up-to-date child development research that probably reflect accurately what is emerging from academe. Scholars, too, are caught up in changing family circumstances. They are asking new questions about child care, about what will or will not help or harm children; and whether more or less parental time, or substitute caretakers, are good or bad. The verdict on what constitutes good family life in the 1980s has not yet been "handed down."

Meanwhile, the messages professional parents hear from their worlds of work are loud and clear: to succeed, your first order of concern must be here. Even though none of this group is employed in the advertising business portrayed in the Academy Awards' best movie of 1979, the expectations and pressures of their environments are comparable to those depicted in *Kramer vs. Kramer*, in the sense that work is supposed to take precedence over a family—invariably.

The role and meaning of work for the second group of people interviewed in this study differs markedly from those described above. For all but one, jobs are less central to their identity than for those in the first group. Work is what they do eight hours a day to make a living; but they convey little sense of excitement, absorption, or pleasure in their jobs per se. The work is not hateful or demeaning; but neither is it challenging or fulfilling. Most of life's interest and satisfactions lie elsewhere.

The Watlings

Unlike the three couples characterized above, the Watlings do not see their work and family lives in conflict, or even difficult to balance satisfactorily.

Gilbert Watling's appearance distinguishes him immediately from the other three men. Like the others, he came to the discussion from work, wearing, in his case, a plaid sport jacket, polyester brown slacks, and square-toed brown shoes. With his playful wit and lighthearted, somewhat cynical view of his civil service job, Gilbert seemed not to take himself or his work as seriously as the others. He laughingly accused the authors of the study of not knowing much about government because the question about hours of work was phrased so ambiguously.

When he handed back his questionnaire he remarked to the distributor, "Do you want to know how many hours we are officially employed, or how many hours we work? Throw out any questionnaires that say they work forty hours a week; I work fifteen hours a week and I do more than most."

In contrast to the work lives described above, Gilbert's federal job in the flexitime agency is much less absorbing, time-consuming and important to him. Although he has risen quickly in the bureaucracy to a high grade (GS 14) for his age (33 years), his background and self-image differ dramatically from those of the three men who were raised in professional families and educated in elite private universities.

Gilbert grew up in a working class family and community in upstate New York. For the past ten years, since graduating from the Merchant Marine Academy and serving briefly at sea, he has worked in the government. He calls his office the "dumb question department." His responsibility is to keep track of nationwide construction in a particular industry, making projections and impact assessments with a computer model.

While Gilbert is proud of his income level and rank in the Civil Service, his primary interest is his extra-work job, namely, building his own house as general contractor. He is as wedded to that enterprise as the lawyers are to the work of their firms. Gilbert describes himself as an outdoor "construction-type" person, mechanically oriented and "happier doing blue-collar than white-collar work." He finds it more rewarding to work for himself than for the government. He contrasts his agency's work to his house building: "It's hard to tell what people are really doing [in government], whether they are contributing to real projects or just paper exercises."

Flexitime is important to Gilbert, largely because it allows him to organize his time around his house building. Essentially he feels flexitime just legalizes what has always gone on. "I have been on flexitime for ten years. The only difference is that now you don't have to watch out about getting caught. Now it's legal.

It takes the stress off—if there ever was any. Having a permanent starting time is a myth for me. I flex every day."

When others in the group said they could not manage supervisory responsibilities with that attitude, Gilbert replied: "I have a supervisory position. They find me if they need me. . . . I don't feel bad about going home at 3:30. The work will be there in the morning. I think that's an attitude [that you have to be there]. They're not going to take the job away from me. I may not get promoted. But I bet I do."

Gilbert's perception of what it takes to rise to the upper levels of his agency is different from both the perception of other federal employees we interviewed with regard to federal employment and from the perception of the private sector attorneys for their field. Furthermore, "making it" means something different for Gilbert. He feels he has risen faster and makes more money than most men his age; he is proud and satisfied with his success.

Gilbert's only objection to flexitime is that it is not fully implemented: "Bosses don't recognize that they're not supposed to call staff meetings except in core hours. There is some policy that could be set down to change it. But nobody will fight the system."

Gilbert's interaction with his sons, ages seven and nine, is largely around the building of the house. Unlike Warren Telford, Ken Garrett, and Steve Benson, he doesn't worry much about how much time he spends with his kids; instead he speaks with pride about their independence and ability to care for themselves. He says only half-jokingly that he made a mistake in going once to a school event: "Now they ask all the time." Commenting on what values his kids have picked up from him about working, Gilbert explained, "My kids are getting a strong work ethic from me; but they are learning to . . . ask why they are doing something."

In the Watling family, tasks are divided along strict sex-role lines, which is as it should be, in Gilbert's view, because his wife "doesn't have to work to support the family." He thinks it is nice for his wife to work, but not essential, whereas for husbands there is no option.

With her warming smile and happy face, Gilbert's wife, Maria, is friendly, self-possessed, and stylish. She came as a teenager from Latin America to the United States to work. Describing her own work choices Maria says, "I looked into various careers and chose nursing because I could choose shifts. I felt it was important to be home with the kids. I didn't want to be cranky. I want to spend time with my children and husband. It works. It's fantastic."

Maria and Gilbert do not feel conflict about their parental and work roles. She explicitly takes primary responsibility for both home chores and child rearing—seeing that her sons do their homework and practice their music before playing—and does not suggest that Gilbert should share these responsibilities more than he does—except to wish that "he would spend more time with the children."

The Caldwells

Loretta Caldwell, a GS 9 secretary in MarAd, came to the discussion without her husband. Six months ago he left his job as an auditor in another federal department in order to open his own business. She talked openly, laughingly, and with cynical good humor about her satisfactions and disappointments. She has short, blond hair and a playful, flirtatious manner. Of all the discussants, she was the most enthusiastic about and grateful for flexitime. But she talked first about the importance of her job in general, especially in the context of an unsatisfactory marriage in which she sees very little of her husband.

> *Loretta*: Last year before going back to work it was just me and the two kids. It was terrible. Such a small sphere. Only one neighbor. My husband not home. It got pretty hairy. Just the kids and no one else to talk

to. . . . But I didn't go back to work because I wanted to. I was told to (by my husband), then I enjoyed it. Now I love to go to work because I have so much fun. It's the people. I have friends at work. I enjoy it. It's much different from cleaning, the kids, and yard work.

I'm not in a real stressful job. In fact, the higher you go up as a government secretary, the less you have to do. I used to have more stress with a different boss. I enjoy working so much I would never want to simply stay home now. I will work till I retire now that the kids are in school full-time. But my oldest child feels the way Louise Garrett's son does. He complains about me working. He would like me to volunteer in school more.

Flexitime is great . . . is good for me. I used to have to rush home and get dinner as fast as I could, now I have an hour before dinner. . . . But flexitime does not help for school and doctor appointments because I can't get to them when I get off work at 3:30 and commute forty-five minutes. Mid-day flexibility is not useful for the kids' things—given my commuting time. It's only good if you work in the environment where you live. Also, I still have to use a babysitter after school. I leave at 6:15 in the morning. My husband is at the house until the kids go to school, or to camp in the summer. He used to work at the Treasury Department, where they had flexitime before us. I don't know what I'd do if we went back to 8:30 to 5:00. I would have to have a sitter one hour in the morning and three hours in the afternoon.

Overshadowing Loretta's enormous enthusiasm for flexitime is her even greater appreciation of the importance of her work in her life, not only for the income (which was a necessity); but more important, for her happiness. In turn, it seems clear that even though the pace keeps her rushing on most days, by her own testimony Loretta feels she is a better mother when employed than when at home full time. And even the modest amount of schedule flexibility permitted in her agency enables her to manage her family life more easily.

In contrast to Loretta's exuberance about the importance of flexitime in helping her balance her job and family life, the professional women in our sample did not praise flexitime as highly for easing the strains. Why is this so? The different feelings about the value of flexitime seem related to several factors. First, to some extent the professional women take schedule flexibility for granted as a norm of professional work. Although they expect to work regular hours, for the most part, their time is not as closely supervised as that of a typical government secretary.

Second, the absorbing nature of some kinds of professional work, even more than supervisory responsibilities per se, often keep professionals psychologically involved in their work even when away from their offices. This "spillover" quality of work interferes with family in the way Ken Garrett described. It is hard to "turn off" one's thinking. The work tends to be a continuum from one context to the next—never wholly left behind in the office the way clerical work is for Loretta.

The O'Haras

Joe O'Hara's intense blue eyes, and lined and expressive face immediately announce that at age forty-two this man has some special life history beyond the government bureaucracy—some knowledge of pain and struggle that sets him apart from the others we interviewed. As a GS 12 in EDA, he currently works on designing and implementing economic development programs for which the agency gives grant and loan monies. In the course of the evening he shares some of the work-related struggles visible in his intriguing face:

> I was in the Seminary for years. I left before being ordained. I have two older brothers who are priests

> and I knew the priest's life. I withdrew because it looked mundane. . . .
>
> I think one of the things I have had to come to grips with in terms of work was a philosophical notion of work—I was taught that work was really kind of the wages of sin. . . . I was looking for a hell of a lot out of work . . . for work to respond to a level of need that it simply couldn't do. I kind of had to bring it down. I don't charge it with as much emotional investment anymore. In the government I think that we're basically about "housework." We're just kind of helping keep things going. There has to be a better world than the bureaucracy—which moves at a snail's pace and doesn't give satisfactions that are concrete. The government is very abstract and not very satisfying. I found out most other jobs are pretty much the same. Once you get away from cotton fields you're into abstraction. And it's difficult to get immediate gratification out of a job.
>
> You get to be aged thirty-five moving in toward forty and the old dragons just ain't worth the fight anymore. All of a sudden, they go away. Now I think of work as a . . . just a necessity of life that one does and just try to make the best of it, . . . not to some sort of standard necessarily . . . but just to the folks one interacts with every day—that's the reality of work, it seems to me; it's not your product as much as the interaction you engage in every day. For the most part I just don't put that much stock in work. As far as my own personal fulfillment, home is my base.

Like the Telfords, home for Joe is a house in surburban northwest Washington—with his wife Kathy, who is a nurse and a professor of nursing, and his two daughters, ages eleven and nine.

Like Joe, Kathy is instantly warm and engaging. But unlike

Joe, she is emotionally involved in her work, clearly exhilarated by the lively interaction she enjoys with students and patients. Her round, cheerful face is friendly and animated, and she talks articulately with enthusiasm about the challenges and pleasures of her job.

Kathy is also very clear about how her flexible hours and proximity to home are important in terms of her being able to balance her work and family life in satisfactory ways:

> I had one year when the children were four and six when I worked a standard eight-to-five job. It was real tough. For lots of reasons I'm not cut out for that type of job—as an administrator. In contrast, aside from liking teaching, there is the flexibility to get up at 4:00 A.M. and grade papers and still be there and get there. . . . I would not like Joe's job. . . . Just Monday to Friday, forty-nine weeks a year; . . . that just doesn't suit me. But that doesn't make my job any better except for me.

Kathy also recognized some job costs in taking advantage of flexibility to be home instead of on the job, in a field where informal aspects of jobs are important, if intangible: "I was getting all my hours in, but I didn't have time to shoot the breeze, have tea, and talk about the day with colleagues. . . . But it is always a balancing thing. They're never clear—my priorities."

Even though the O'Haras do not think of themselves as unusually stressed in balancing work and family responsibilities, their description of the competition between their work and family demands illustrates a conceptualization of the links between work and emotional life that was developed in another interview study of family life (Piotrkowski 1979b). Piotrkowski views the connection between work and family in terms of systematic competition for time, space, personal availability, and energy. The necessity to sustain life by time and energy sold outside the home causes a depletion of the resources necessary to achieve familial goals of intimacy and meaning. In turn, the systematic conflicts between work and family systems are expressed as strains, or stress, like those described by the O'Haras. Similarly, patterns of separateness and connectedness are influenced by physical absence and its timing—and the abilities of family members to cross boundaries between the work place and home.

Kathy and Joe O'Hara further illustrate Piotrkowski's formulation by expanding on their complications in keeping family life functioning smoothly.

> *Kathy*: I was slow to realize that Joe and I can go along pretty well as long as we both don't hit a crunch at the same time. But then we really get into trouble at home because there isn't anybody who has the extra time to fill in. It's awful. It gets very tense. It's hard on the kids. All the things that are stable in our lives, like an evening meal, go out the door because nobody's home to fix it and it's havoc. Now that it has happened a few times we know what was wrong. But I don't know how to guard against it.
>
> When things are really tense there comes a competition for time. I can say on a given day or week to Joe that I am going a hundred and fifty miles an hour next week and he sort of picks up the slack. He is home early, does more stuff. Or he can say he will be out of town and I will do more. But when we both need that time and it is not there—it just isn't good. Then you live through it and you put the pieces together afterwards, but it takes a toll. Because our work lives are really not all that flexible, there are times when it really has to be heavy for both of us. Thank God it is not all that often.
>
> *Joe*: [Nodding head] I think the first thing that happens is that things get out of sync. It is not as if you can look ahead a week and say, you're going to be busy and I am going to be busy, so we arrange that we don't get tense. No, No! Usually it is tense first and then . . .

> *If you were going to try to fix things so that wouldn't happen, is there a way?*
>
> *Kathy:* I don't know if there necessarily is a way to fix things. The only way is to learn to live with it . . . it helps me to understand what is happening.

When asked if flexibile work schedules would help when things get "out of sync," Joe, who works in the standard time agency, named a different factor affecting the interconnections between his work and family life as being an even more important constraint on his flexibility:

> *Joe:* My inflexibility is transportation. I'm either a car ride or a bus ride from home. It [transportation] is less flexible than the government regulations about leave time. I've never been in a situation where it has been difficult to get leave from a supervisor.
>
> *Kathy:* Yes, proximity to work is important. The farther away you are the more distant everything is. . . . The more distant it is, the more it becomes two separate worlds.

Like the more job-absorbed couples described first, the O'Haras do not have many suggestions about how to ease the complications of balancing their work and family needs. While Joe does not seem as worried as Ken Garrett and Warren Telford that he may be making wrong choices, he also seems to accept that the syndrome is immutable.

The O'Haras' relative lack of pressure, in contrast to the other couples, may be explained by two factors. First, Joe and Kathy's total employment and commuting hours are twenty hours less than the average of the other three couples. In the other cases, the longer work hours are those of the Garrett, Telford, and Benson men. Second, Joe O'Hara—about five years older than the lawyers and ten years older than the "budgeteer"—is nowhere nearly as work-absorbed as the other three men.

When asked if he or Kathy were interchangeable in terms of doing things at home, whether he can do everything she can do, Joe replied, "Yes. But I don't." In this connection, regarding their roles at home, the O'Haras talked more about the differences in male-female roles.

> *Kathy*: It's the basic responsibility, who plans the children's day. What happens when you're going to be three hours late? You have to get a babysitter, or take the kids, or explain to the kids.
>
> Joe goes every morning at a quarter to eight and 90 percent of the time arrives back at six. That's just a block of time that is Daddy's work! Mine's much more . . . well, I'll give you an example. I was at work this summer at the hospital. The swing broke. I could hear "Wha-wha" on the phone. I hopped in the car and drove home. I put on the bandaid, walked her over to the school yard, and was back at work in forty minutes. . . . Part of that story has to do with my having a car. But it is also the role I have assumed, and so they [the children] are much more a part [of my work].
>
> *Joe*: Kathy, let me ask you a question. Suppose I had been the one who had gotten called, and I called you and said I was on my way to take the kid to the doctor. Would you feel comfortable about staying at work while I took the kid to the doctor?
>
> *Kathy*: [Laughing] My fantasy, and this is probably not fair to you, is that you would say to the crying kid, "Just wait an hour and call me back." . . . It's a hard question to get a handle on. I don't think I have to do it (family work) all by myself . . . but I signed up for it.
>
> *If your daughter has children, does it mean she will "sign up" for what you do, as opposed to what Joe does?*
>
> *Kathy*: I think yes. But we—Joe and I—have come a long way. These children start with a father who does the dishes every night, irons his own shirt in the

> morning, takes care of them if their mother goes off to St. Louis for three days. They're starting farther down the pike than I certainly did.

Kathy O'Hara's account echoes the Garretts' picture, not only in terms of which parent takes the lead in daily family responsibilities, but with regard to what the children's expectations for parents' roles are—as a reflection of what they are used to.

The Petersons

In addition to whatever natural shyness characterizes Karen Peterson, her nearly inaudible soft-spokenness and tentativeness in expressing her own views also seem a function of her role in her marriage as a deferential wife—and her newness in her job. For six months she has worked two days a week as a secretary in a small, quiet office with no pressure. Karen's mother, who lives outside Boston, tells Karen in regular telephone calls that she opposes her working; she thinks Karen has too much to do around the house. For Alfred Peterson, like Gilbert Watling, it is important that Karen does not have to work; when she first talked about wanting to work, he wanted to know why.

For the time being, Karen feels she has no difficulty balancing a family and work; in fact, she sees her current work demands as lighter than the load she carried when doing volunteer work for her church and the cub scouts. She thinks that her job makes her kids appreciate her more. Even though she feels she got more direct satisfaction from the one-to-one involvement with patients in her volunteer hospital work, she feels more important and respected as an employed woman.

In the twenty years her husband has worked in the Maritime Administration, Karen has never been to his office—nor have their three children, two of whom are in college. Alfred, a GS 15, likes to keep his work, as a supervisory engineer, and his family life separate.

Except for slight lines of maturity in his smooth face, a snapshot of Alfred Peterson today at age forty-two would probably duplicate his undergraduate yearbook photograph with his 1959 short, neat blonde officer's haircut and firm jaw line. Alfred speaks evenly and quietly, taking long draws on a thin brown pipe between his comments. He mused, as follows, about the connections between his work and family life:

> I don't even think my kids know where I work—except for my oldest son who's in the same type of work now. But, I don't think they know anything about my type of work. As a matter of fact, I don't think you [looking at his wife] know much about my work. . . . I don't talk about work at home, I keep it almost entirely separate. It's a different kind of life for me at work. . . . I know I have to get things done through people and so I adopt a working style which is entirely different from my life at home. . . . At home, there is not the push to get things done like there is at work. And the other thing, I guess, is my work is at a desk and when I go home I want to forget about the desk. I'd rather do things with my hands . . . build houses, that kind of thing. I guess that's why, really, we bought summer property and have been building a house. It's taken eight years and it's almost done.

In contrast to Gilbert Watling's enthusiasm for flexitime in his life, Alfred Peterson feels relatively blase and indifferent towards it:

> *Alfred*: I rely on an 8:00–4:30 type of job. I like that because I plan around it. . . . Maybe it's just because with my engineering background and training I like to put things in little slots.
>
> *When you got flexitime in MarAd did that jar your orderly sense of things?*
>
> *Alfred*: No, I don't use (flexitime) that often. Only

> when I have to. I adjusted my work hours by half an hour only because I carpool. If I'm an hour late I take an hour of annual leave.
>
> *What if a man working for you was four hours late for the kind of family reason Joe and Kathy were describing—that they both had incredibly complex demands that week. What is your reaction as a supervisor?*
>
> *Alfred*: That doesn't get me hyper at all.
>
> *Karen*: [Quickly interjecting] Because they would be taking off annual leave, meaning that they have to cut into their own vacation time.

But the issues could arise with more conflicts in another kind of work setting, as Kathy O'Hara insists:

> *Kathy*: I don't, God forgive me, consider either one of your [Joe's and Alfred's] schedules all that tight, compared to being a nurse on a floor where you're covering people sick and dying. Suppose you get a call that your kid was hit by a car; it gets complicated to leave. The total hours are one thing; but how much flexibility there is in the eight hours on certain jobs is another.
>
> *Joe:* The only place real pressure happens in the federal government is high up—if you're an assistant secretary or a deputy or a political appointee or a GS 17. . . . At the level of folks I know, nothing prevents me from getting up and leaving when I want to.
>
> *Alfred:* I think you retain the flexibility at higher levels too. You just put in more time. It's the same with travel. I can schedule it when and as much as I want. . . . As Joe says, the bureaucracy is not a crisis all the time. It is not like working in a studio and if a key person is off the show does not go on. It is not that kind of situation at all.

Alfred's comments echoed Gilbert Watling's in two senses. One, both men felt little external work pressure in their sector and level of the federal bureaucracy; and second, neither of them had the kind of internalized work drive of Garrett, Telford and Benson.

With fewer people in the second discussion group, we had time to ask what kinds of work policies they thought would make a work atmosphere supportive of families. Kathy O'Hara had a clear view of the important issues. Karen Peterson said nothing. And Joe O'Hara and Alfred Peterson felt strongly that women were still the best child raisers and that employers had no responsibility for accommodating these responsibilities.

> *Kathy:* When you're married with kids you basically need respect on your job for the fact that your family is a priority in your life, respect that can hear you when you have got to go early or come late. People should not be penalized for having families, for example, women taking leave for babies. . . . The easiest people for me to work with are those who have shared my experiences.
>
> *If Alfred had paternity leave when his son David was born would that have made him a different kind of boss?*

When Alfred was asked this question his quiet, soft-spoken wife Karen smiled broadly and giggled. Kathy answered: "I sure would put 50 cents on it. Alfred would have been humanized. If anyone is pulled out of one single part of their life and has to plug into another—not forever and because they choose to do it—it can be enriching. He would go back a broader person because of it."

But Alfred looked puzzled as she spoke, and did not follow Kathy's reasoning:

> *Alfred:* What is the rationale for parenting leave? That the child would benefit?
>
> *Kathy:* That the father would benefit, too. I would want Joe to stay home some but not in the place of me.
>
> *Alfred:* What responsibility does society have to sup-

port something like this? It sounds incredible to me.

Joe: If this idea was enhancing, if it meant less family breakup . . . I wonder: can one get at the nut of solid family life in this way? Are we going towards a system in which we can set up an environment where life will flourish, one which would be conducive to family life?

Kathy: Does work need to change or does where men "come from" need to change? I don't know how much work can lead it.

Joe: The point is we don't know. To launch a full public policy on family life is a dangerous issue.

Alfred: I can see working mothers having leave for a sick child.

Joe: Me, too. If you had a policy that said women with young children . . .

Kathy: Women *or* men.

Joe: I am saying women . . . Watching you and my mother, one working women's generation over forty-one years . . . my impression is . . . there's the element of motherhood that plays into it, . . . a mother is more quick to respond in those kind of things than I would be.

Kathy: And by God, you're not going to change it, are you?

Clearly, in all our interview families, fathers and mothers play different parenting roles from one another; in all cases the mothers still carry the major daily parenting responsibility. The women feel they "signed up for it," meaning primary parenting, as Kathy put it. The mothers, however, would also like to have more help from the other parent. Our survey data on this point suggest Kathy is typical of the women in our sample. On the question which asked mothers whether they would like their husbands to spend more time with their children, almost 70 percent said yes; and more than 40 percent wanted more help with home chores from their husbands. In contrast, only about 15 percent of the men said they wanted their wives to spend more time with the children or on chores.

Sympathetic and sharing as Joe is about family work, he says bluntly that he also is not ready to split things evenly. In part, Joe, Alfred, and Gilbert think mothers still have some special things to offer children, particularly pre-schoolers, that fathers do not. For example, Joe says, "Women do a thing with babies that is different from what men do. . . . But even if you give women child leave up through ages three to four, there's still a question of whether the public should pay for it. These women should be allowed to stay with kids as long as society wants. Then women can move into the work force if they want."

In Ken Garrett's and Steven Benson's cases, the rationale for doing less than half the family work is something else, namely, that the demands and rewards of work prevent it. But the net effect is the same. The women in our study, like those in all the time-budget studies, bear a disproportionate share of family work and they feel the related stress.

For this family impact analysis the important point is that the pattern is true irrespective of several key work factors:

- The absorption of the work.
- The level of responsibility.
- The pressure of the job.
- Whether or not either or both spouses have flexible work schedules.

As with the survey results, however, it is essential to emphasize that almost all respondents like schedule flexibility. It is valued by the women even more than by the men. Yet, in and of itself, flexitime does not have enough "family impact" to affect the major family factors and functions considered in this study—namely, to reduce stress related to the overlapping demands of work and family, to increase male family work, or to increase equity between men and women in family work. Much more important than schedules in affecting how people balance work and family roles are two other factors: first, what men and women each feel most responsible for in work and family; and second, how absorbed they are in their work and why (e.g., internal and external work demands).

9

Study Conclusions and Policy Implications

GIVEN the increasing use of flexible work schedules and the claims for their value to families, this study of federal employees on flexitime aimed: first, to contribute empirical data to policy and research on the topic; and second, to experiment with the process of conducting family impact analysis on public policies.

As stated above, the project appeared simple. A particular workplace innovation of recent years had caught the popular imagination. It seemed to please both employers and employees—and to be of benefit to people's lives beyond the workplace. Observers from many quarters enthusiastically proclaimed that flexible work schedules are good for family life.

Investigating these claims for flexitime required looking behind the cheerful consensus to find out, first, what improvements in family life people expected from flexitime—and for which family members; and second, what differences in family life, if any, there were between a group of workers on flexitime and a group on standard time. Finally, the outcomes of these two inquiries suggested questions about what additional social policies might be useful in helping people balance job and family responsibilities.

The results of this investigation challenge the predominant optimism about flexitime in five important ways:

- Analysis of the testimony in the congressional hearings on alternate work schedules revealed that various spokespersons had virtually contradictory expectations for how flexitime might help families.
- The survey data in this study revealed that the families most helped by a modest flexitime program are those with the fewest work-family conflicts, namely, those without children.
- More flexibility in work schedules, as well as other programs and policies, probably are required to help families with the most pressing work-family conflicts.
- The supplementary interviews suggest that complicated and unresolved value questions both about men's and women's roles, and about the relative importance of work and a family, underlie the ambiguous expectations for, and effects of, flexitime.
- Minor changes in the formal conditions of work (like scheduling) will not significantly affect the family variables measured in this study (like sharing of family work). More substantial structural changes, as well as shifts in values and expectations about people's participation both in work and a family must occur first.

Contradictory Expectations for Flexitime

With respect to the hopes of how flexitime may help families, the contradictory expectations appear to have emerged in the following way. At first glance flexitime seems a logical way out of the most frequently mentioned work-family conflict, namely, between children's need for parental time and parents' need for work time. Flexitime seems to respond directly to the incongruence between children's school schedules and adult work schedules. In addition to the traditional conflicts between work and school schedules, other work conflicts—for people with or without school children—include those between the demands of the job and the needs or desires of workers for time alone; time with children, friends, spouses, and other relatives; or time in community activities, recreation, or other interests.

By age three, most American children are away from home part of the day, but usually not as long as their employed parents. If school and work schedules were better synchronized, the logistical conflicts between jobs and schools might be reduced for employed parents. Allowing parents choice in scheduling their work might allow them to coordinate their work-family responsibilities more easily—and permit them to spend more time at home.

But a few pieces of the logistical puzzle are left over. Most workdays are still several hours longer than school days. What should be done with children under age three for whom out-of-home care is not universally available? What should be done when older children are sick and cannot go to school? Or when elderly or infirm family members need attention? How can job demands and schedules accommodate all the activities related to sustaining family life that take place outside job and school hours?

As soon as these questions about family care become concrete, the notions about how flexitime will help families become elusive. Who (meaning men and women), will do what (meaning take care of dependents and household chores), and when (meaning parts of the days or years)? Where will the care take place (at home or elsewhere)? Or what will be gained (meaning satisfactions in time with children, grandparents, and domestic activities) by those who take time from employment (or leisure) to give to family work? Also, who will pay for the time spent in child rearing? Individual families or general taxation? Private or public systems? When both men and women are employed in comparable ways, in terms of hours and responsibilities, can flexible work scheduling affect how and by whom these functions are managed?

Conflicting answers to these questions emerge from the congressional hearings on alternate work schedules. In this sense, the focus on flexitime may be seen as a benign battleground for an historically novel work-family debate: should males and females have equal roles in both? Within the comfortable coalition supporting flexible work arrangements are two competing points of view about male and female roles in work and family.

The first view assumes that flexitime, and other alternate work arrangements, are useful in order to allow women to continue to bear primary responsibility for family work—with less stress. This outlook essentially expects contemporary women to embody both the female-family ideals of one era, and the male-employment ideals of another; that is, women are expected to perform both at-home family functions and at-work job functions as if neither of the other responsibilities existed. At the same time, reverse expectations do not prevail for males. They are not expected to take primary, or even equal shares of domestic responsibilities. Thus, imbalances between males and females are the expected norms.

The second view of male and female work and family roles holds that men and women should have equal employment and family opportunities and responsibilities; flexitime and other alternative work arrangements should be used by men, as well as by women, for this purpose.

What seems unique in this view, in contrast to earlier historical periods (see Chapter 2), is the expectation that the labor of men and women should be interchangeable; that is, either a man or a woman should be able, and should be eligible, to do any job and any family activity (except giving birth to and nursing babies). In other words, the division of labor based on sex should fall away both at home and at work.[1] Thus, the contradictory expectations for flexitime flow directly from the competing contemporary views of appropriate male and female roles in work and family.

Employed Parents Need More Than Flexitime

The survey findings suggest that the families most helped by flexitime are those with the fewest work-family conflicts, on the

[1]The opposing view that women are biologically more suited to rearing chil-

basis of the following reasoning. Employed women in the sample continue to bear primary responsibility for family work, as measured by hours spent on home chores and child rearing, whether they or their husbands are on flexitime or not. Women without children have less stress if they are on flexitime; but mothers (i.e., those workers with the most job-family conflicts), with or without spouses, are equally stressed on either work schedule. Thus, the first point of view above—that women should still bear primary responsibility for family work—is sustained. But those women presumed to have the most to do—i.e., mothers—have no less stress when they have this modest flexitime option.

With respect to the second view above, men on flexitime with employed wives do not spend significantly more time on family work than those on standard time whose wives are also employed. But flexitime men whose wives are not employed do spend more time on family work than their standard time counterparts—and they feel less stress.

What can be made of these somewhat confusing trends in stress and the division of domestic responsibilities in the survey data? To those who hoped that employed parents will have less stress, will spend more time with their children, and will share family work more equally, the flexitime program examined in this study seems to have little to offer. Yet recognizing that such social changes are not likely to occur overnight (e.g., within a year of the initiation of the MarAd flexitime program), some encouragement may be found in the fact that fathers on flexitime whose wives are not employed are spending more time on home chores and with their children than their standard time counterparts (but not more time than the two-earner fathers on both schedules whose wives are employed). This trend prevails even when controlling for the fact that the standard time men

---

dren than men has received new support in Rossi's recent biosocial research (1977); she argues that by virtue of childbearing, women have innate capacities for nurturant behavior that men can only be taught.

work an average of two hours a week more than the flexitime group.

Thus, for three groups—fathers with unemployed wives, employed married women without children, and all single people—the MarAd flexitime program makes a measurable difference in the ease with which workers can take care of personal and family chores and activities. In short, for people without primary child care responsibilities, a slightly greater degree of control over their time helps a lot.[2]

For those employed adults with the most family-related obligations, however (namely, single mothers and employed parents with employed spouses), the small degree of schedule latitude permitted in the MarAd program does not make a measurable difference in job-family stress, or in the amount of time workers spend on family activities—suggesting, therefore, the necessity to look further than flexitime for ways to help such families.[3]

Sex-role Expectations Create Confusion

The interviews with a small group of survey respondents and their spouses enrich and enliven the information from the sur-

[2]But there are two additional groups of people without children who do not seem to be much helped by flexitime, namely both one- and two-earner men without children. They spend about three hours more a week on home chores than their standard-time counterparts (the difference is not statistically significant); but they do not have significantly less stress. For them, the slightly greater time on chores possibly explains the greater stress. On the other hand, for the two-earner women without children, who spend two hours a week more on chores and feel less stress, the reduced feeling of stress may be explained in terms of sex-role expectations (discussed in Chapter 4 in reference to the Management Scale). In short, the men who do more home chores may feel more stress because they are not used to those responsibilities; whereas women who normally expect to carry the major home management and coordination responsibility may feel genuinely less stressed when they can leave work an hour or so earlier.

[3]As Joseph Pleck and his co-authors have pointed out (Sept. 1978, p. 7), a

vey in several important ways. Above all, they suggest that as desirable as flexible work schedules are for most people, other factors are far more influential in determining how people distribute their time and energy between jobs and family life. The most powerful influences are sex-role expectations and work expectations—both internalized and institutionalized.

Like the survey respondents at large, the interviewees had different expectations for the work and family roles of men and women. Even when career training was the same, career performance was different: women expected less of themselves in quantity and sustained quality. In turn, they were more absorbed in their family lives than their husbands. Even men who wished to be more engaged in child rearing found it difficult because of their work pressures and commitments. And men who were less work absorbed simply felt less responsibility for day-to-day home and child care—and in turn their wives did not expect it from them.

In short, cultural values about work and family roles for men and women dictated who did what, as well as what stresses were felt by whom—more than the formal work structures like flexible scheduling. For professional men and women, in particular, the current asymmetry between male and female roles in work and family, plus the heavy time requirements of each, leave enormous uncertainties about where their energies and emphases should be. Flexitime helps them to juggle the choices around the margins, but it does not help decide how to make the choices. For less work-absorbed women with family responsibilities, a little schedule leeway is a great boon in managing the logistics of daily work and family life—but these women are still left with the primary responsibility for most of the home demands.

---

large part of what has been recently characterized as "work-family conflict" is really work-parenting conflicts. Labeling the problem as one involving specifically parenthood may help focus public attention on the problem in a more concrete way than the usual formulation.

New Work Policies Needed for Family Well-being

Perhaps the major importance of the present study for policy-makers and employers is to dramatize the fact that families with the most work-family conflicts may need more accommodations to their dual sets of responsibilities than are currently available. In other words, in light of changing employment, fertility, and life expectancy patterns, even stable, middle-class, civil servant parents—like those in the study—may require more substantial changes in the structure of work than minor schedule flexibility if they are to better balance their lives both as employees and family members.

Eli Ginsberg, chairman of the National Commission for Employment Policy, has stated his sense of the changes sweepingly: "A revolution that alters the relation of one sex to the world of work will inevitably impinge on a great many social institutions and mechanisms. . . . No aspect of life will be untouched by the revolution in womanpower, and there is reason to believe that the changes will improve the lives of both sexes" (Ginzberg 1975, p. 154). Similarly, in Kamerman's stark view, failure to respond to these changes will have negative and predictable consequences for families:

> Clearly unless women are supported in their efforts at fulfilling work and family roles, one of these domains will suffer; they will perform inadequately at work and remain (if at all) in low wage, low status positions; they will have fewer children, or none; they will have children but be inadequate as mothers; family crises will occur with frequency.
>
> Kamerman 1978b, p. 35

The popularity of flexitime results, in part, from the fact that it seems to be an easy structural solution to the strains produced by rapid social changes. But its very simplicity obfuscates the competing hopes for what it can accomplish. On the one hand, flexitime is agreeable because it costs very little and changes very little about what employers can expect by way of a day's work from employees—while apparently making people more cheerful about what they get paid to do.

On the other hand, for those who want to alter the traditions which make it difficult to combine "paid work" and "family work," the concept of self-determined working hours seems a step towards a multitude of far-reaching changes in the structure of work. But on the basis of the findings in this survey, modest versions of flexitime alone—like that in the Maritime Administration—cannot be considered significant changes in the existing organization of work, that is, in terms of their impact on work-family conflicts. The complexities of the social, economic, and demographic changes that now encourage men and women to have comparable work roles throughout their child-rearing years require policy responses far beyond flexitime—if the goals for families with children are, in part, those held out for flexitime, namely, to reduce parental job-family stress, to enable parents to spend more time with their children, and to increase equity between males and females in paid work and family work. What other policies might help? At least two levels of change can be discussed: expansion of existing programs and large-scale economic and land-use changes.

Expansion of Existing Programs

Those which are already in use and could be used more widely in the United States include:

- Parental work-leave policies. In Sweden parents may divide nine months of leave after the birth of a baby. Either parent may stay home with the infant. France and Norway also have national parental leave policies.
- More flexible work schedules and flexiplace. New United States federal flexitime legislation allows people to average eighty hours every two weeks, with more or less than eight

hours daily, as they wish (within strictures of agencies who have made use of the new law). Some businesses now allow mothers on maternity leave to work at home, as far as possible in terms of their assignments (Kronholz 1978). Sabbatical leaves are being conceived in terms of redistributing time to education, leisure, family work and paid work over the life cycle (e.g., Best and Stern 1977; Strober 1976).

- Shorter work days for parents (i.e., variations on part-time). Parents may now work six-hour days in Sweden until their youngest child is eight years old, with less pay than for eight-hour days but without jeopardy to their jobs (Ministry of Health and Social Affairs 1977).
- Increased pre-primary school programs and more after school programs. More programs for children from age two to three through age five are desired and in use in some places, for example, in Belgium and France. Expansion of such programs will gradually do away with the artificial historical dichotomy between day care programs and pre-primary care, and will acknowledge the interest of most parents in having their children have some group experience from ages three to five (Kamerman 1980; Kamerman and Kahn 1979, 1980).
- Parental insurance (instead of maternity benefits). Swedish parents may divide between them an entitlement to stay home from work for child rearing, either as an absence of two to four hours per day, or for a period corresponding to three months of full-time absence from work at any time up until the child is eight years of age. They will be paid at 90 percent their actual income for two of the three months (Ministry of Health and Social Affairs 1977).

Economic and Land-use Changes

More utopian and costly efforts to respond to work-family conflicts might include attempts to reduce the major physical separations between work areas and living areas that now characterize American city and suburban life. Energy shortages may already be encouraging families to try to live closer to their workplaces. But without larger upheavals in existing job, housing, and school patterns, the physical and temporal distances between family life and work life cannot be substantially reduced for many Americans. But perhaps, as suggested by Myrdal and Klein (1956), policymakers should increase their efforts to reverse the trends that have functionally divided huge metropolitan areas.

Value Changes Needed for Better Work-Family Balance

Even if policies and programs like those suggested above are legislated, they will not alleviate work-family conflicts unless they are accompanied by shifts in values about the appropriate connections between work and family life, as well. Kanter's list of the five dimensions of work experience that have bearing on family relations (see Chapter 4) suggest areas in which prevailing values may have to change in order to mitigate work-family conflicts.

For example, flexibility in job schedules will increase time in family work only if people have jobs in which they can earn enough to support their families without excessive overtime; if the jobs are challenging and fulfilling; and if they choose to spend some of the non-work time in family life. Excessive work hours will be reduced for committed professionals only if their criteria and rewards for successful lives are not synonymous with long work hours. More equal sharing of family work between spouses will occur only if expectations about sex role divisions of paid labor and home labor are altered.

Demographic and economic factors are also likely to influence each of the above. For example, as people have fewer children and longer work lives, more time with the children may seem more desirable. As women share more of the economic support of families, and men are encouraged to value their family time more, men may feel less compelled to be absorbed in work. And as women's lives are challenged and diversified by participation in the labor force, they may have greater pleasure and success in their parenting.

To conclude, as this study does, that flexible work schedules of the sort examined in this research cannot alone make significant differences in the work-family conflicts of employed parents is not to discount the potential importance of an innovation like flexible schedules. On the contrary, unlike the more economically and politically difficult programs and policies suggested above, flexitime is being used—and being expanded conceptually—as in the new experiments for federal workers authorized by the September 1978 legislation. Flexitime may be an important first step towards altering the traditions in which work roles and work organizations have been defined and structured as if the family did not exist, as the Rapoports and their co-authors put it,

> except in the background, after hours and compliant to the primacy of the work role. . . . What is needed is a more open-minded, innovative approach to the problem of the structure of work. . . . It is difficult for individuals or families to change a work situation. Though many . . . may wish to accommodate family interests, they may also feel uncomfortable or insecure about being the one to be pushing for change. . . . At present, a frequent situation is one in which husbands indicate a readiness to be "fair" about taking domestic responsibilities but feel reluctant to ask employers for latitude to do so for fear of losing ground in the competitive situation of the job . . . The idea that workers should be primarily responsive to the job is so ingrained that it is impossible for many employers to envision change. There may, therefore, be an actual collusion between employers and employees to see any alteration of the status quo as out of the question.
>
> Rapoport et al. 1976, p. 178

Pleck also sees the problems of contemporary families as extending beyond individuals and requiring institutional and value changes:

> It does not seem possible for large numbers of families to function with *both* partners following the traditional male work model. Such a pattern could become widespread only if fertility dropped significantly further or if household work and child care services became inexpensive, widely available, and socially acceptable on a scale hitherto unknown. In the absence of such developments, greater equality in the sharing of work and family roles by women and men will ultimately require the development of a new model of the work role and new model for the boundary between work and the family which gives higher priority to family needs.
>
> Pleck 1977b, p. 425

Afterword

Evaluating Family Impact Analysis

THE process of family impact analysis, as developed by the Family Impact Seminar and applied in this study, is more of a generic outlook, or approach, than it is either a theory or methodology. The purpose of the analysis is to encourage policymakers, scholars, and citizens to evaluate public and private policies and programs in terms of the range of intentional, or unintentional, effects they may have on people's family lives. The first chapter of the book outlines the process by which the topic of this study became a subject for such an examination; the next eight chapters describe the exercise and its findings. This afterword concludes the discussion by reviewing our experience with family impact analysis in terms of its utility and its limitations for future policy analysis and research.

The present experiment in "family impact analysis" represents a stage mid-way between the earliest efforts to define a process (e.g., Kamerman 1976; Mattessich 1977; Wilson and McDonald 1977) and more recent efforts to delineate appropriate steps (Ory and Leik 1978; Family Impact Seminar 1979). The newest documents, from the University of Minnesota Family Study Center and the Family Impact Seminar at the George Washington University, lay out two ways to undertake a family impact analysis of a particular policy. Although they vary in emphases, the two guides are complementary; each tends to emphasize the lessons from the specific experiments with which their process was developed.

Our study of work schedules effects on families has left us sobered by the difficulty of the process, but hopeful about the value and feasibility of governments or other groups conducting such analyses of public policies. On the one hand, the conceptual and measurement problems are so elusive and complex that it seems impossible to assess, in a comprehensive and exact way, the family impact of even a few of the multitude of policies that touch American families. On the other hand, to ignore the ways in which work practices and other policies affect contemporary life, simply because the process is complicated and imprecise, is even more irresponsible for those concerned with the set of social problems currently collected under the rubric of family issues. The following discussion suggests some of the ways in which the evolving frameworks of the Family Impact Seminar were helpful in conducting family impact analysis of flexitime, as well as some of the limitations of the approach and of the design of this study. Interwoven into the discussion are suggestions for future family impact analyses and further research—with the distinctions between the two necessarily being imprecise.

The frameworks for family impact analysis illustrated here were originally presented and described in detail in the Family Impact Seminar's Interim Report (1978b). They provided a useful matrix of questions for considering how work-scheduling policies could affect the lives of government employees and their families. On each figure we have checked the boxes indicating the kind of information we tried to take account of in designing the survey and the various interviews with policymakers and families.

The public policy framework (Figure 10) alerted us to examine the laws permitting the establishment of flexible work schedules (i.e., the interpretations of the old statute by the Civil Service Commission in 1974, and the passage of laws in September 1978

FIGURE 10
EVOLVING FRAMEWORK: PUBLIC POLICY DIMENSIONS

Implementation Components	Value Assumptions	Levels of Government: Federal	State	Local
Historical background	■	■		
Laws: Act(s), amendments, court interpretations	■	■		
Regulations		■		
Appropriations: Funding levels, allocations, terms (incentives, disincentives)				
Administrative practices: Standard procedures, guidelines	■	■		
Implementation characteristics				
• Auspices: private/public (schools, hospital, agency, workplace)		■		
• Staffing: professional/ bureaucratic (orientation, training, affiliations, unions)		■		
• Convenience and accessibility to families (hours, location)		■		
• Coordination (with other programs)		■		
• Sensitivity to families' needs and realities		■		
• Nature of relationship with family		■		
Related programs/policies		■		
Related laws and court decisions		■		

to extend flexible scheduling experiments); the regulations establishing specifications, such as the core times (when everyone must be at work); and administrative practices, such as whether a particular supervisor supports or undermines employees' use of flexitime, or whether employees tend to "cheat" on the system.

In addition, Figure 10 encouraged us to look at other family-related work policies, such as part-time employment, on-site child care, transportation facilities, and parenting-leave policies (no. 6 in the implementation components).

The family impact framework (Figure 11) facilitated our investigation of work schedules in the following ways. It led us to consider the economic, socialization, and nurturant functions of the family in relation to work schedules. Most important for our focus on scheduling policies, the framework emphasized the coordinating and mediating roles of workers in family life; for example, whether flexitime seemed to enhance the ability of a working parent to manage chores, errands, and appointments related to the functioning of the family unit.

Furthermore, the framework ensured attention to family socioeconomic characteristics (e.g., occupational and educational level—no. 1); family structures (e.g., numbers of parents and earners in the family, etc.—no. 2); and the family life cycle stage (e.g., with or without children or grandparents at home, age of children, etc.—no. 3). In terms of immediate family contexts, the framework directed attention to the effects of flexitime on internal family relationships, in particular, on family stress and family work (i.e., parental time on home chores with children—no. 4).

Our survey made less extensive use of the other three horizontal rows on Figure 11, namely, the pluralistic context (e.g., ethnic and racial factors—no. 5) and the informal social network dimension (e.g., the extended family, community groups, etc.—no. 6). In the neighborhood environment dimension (no. 7), we asked only about transportation for commuters in terms of time; and in retrospect, we think perhaps we should have asked more

FIGURE 11
EVOLVING FRAMEWORK: FAMILY IMPACT DIMENSIONS

Family Types and Immediate Contexts	Family Functions		
	Membership Functions	Economic Support and Consumer Functions	Socializing and Nurturant Functions
	Coordinating and Mediating Roles		
Family types			
Socio-economic characteristics: income/occupation/education		■	■
Structure		■	■
• Single parent/two parent			
• Nuclear/extended			
• None/one/two wage earner			
• Orientation/procreation*			
• Primary/reconstituted†			
• "De facto" families‡			
Life cycle stages		■	■
• Early family formation			
• Family with school-age children			
• With children in transition to adulthood			
• With no child dependents			
• With elderly dependents			
• Aging families			
Families' immediate contexts			
Internal family relationships		■	■
• Interdependency: economic/psychological			
• Conflicting/complementary rights and interests			
Pluralistic context: ethnic/religious/racial/cultural values and behavior			

Family Types and Immediate Contexts	Family Functions		
	Membership Functions	Economic Support and Consumer Functions	Socializing and Nurturant Functions
	Coordinating and Mediating Roles		
Informal social network: friends, extended family, neighbors, community groups		■	■
Neighborhood environment: housing, commercial transportation, recreation, municipal services		■	■
Sex role attitudes or satisfaction		■	■

*Family into which one is born (orientation), and family that one creates (procreation).

†Families of first, second, or subsequent marriage.

‡Families that are not defined by blood, marriage, or legal adoption (formal or informal foster and informal adoption families).

questions on this topic—for example, by what means people commute and how problematic commuting is—because these factors appear to interact strongly with the use of, and attitudes towards, flexitime.

We also added two additional dimensions to the framework for our study of work schedules: first, sex role attitudes (e.g., what respondents think males and females ideally should do in terms of work and family); and, second, work attitudes (e.g., whether money or other factors are more important in work, job satisfaction, and the like). (But these additional questions were not analyzed for this report of the study.)

Thus, on the positive side, the family impact analysis framework provided a comprehensive guide to considering the multiple effects of programs on multiple aspects of family life; in other words, it encouraged an "ecological" approach to assessing the effects of a policy on family life. In terms of work schedules, in particular, the framework illustrates how to bring family needs to the attention of employers and demonstrates how to alert policymakers to the family impact of policies established with other major purposes. On the other side, our use of the Family Impact Seminar frameworks revealed that the existing formats had several limitations with regard to this type of study, and pointed to several ways in which they could be further developed.

In their present form, the major value of the frameworks is breadth, not depth; that is, they can help policymakers, researchers, or citizen groups to think about the wide range of influences on family life; they can also suggest important questions and indicate the many stages between the creation of a law—or policy—and the ways it may touch families' lives. Similarly, the ecological theory of Bronfenbrenner and the family systems theories, which provided our overall conceptual approaches, require attention to the links between mesosystems (e.g., work structures) and microsystems (e.g., families).

For both family impact analysis and family research, the most important and far-reaching need is for further development of theory. The ecological and systems theories indicate that there are reciprocal connections between work and family experiences, but they do not yet specify what those influences may be. While the recognition of a relationship between work and family is a requisite step, additional theory is needed to explain which elements of work influence which elements of family life, and in what ways; in other words, how the relationship operates. Theoretical advances on this topic will need conceptual models that can hypothesize work-family relationships more concretely, in terms of systematic interactions, and can identify the aspects of family life to measure in connection with particular policies.

For example, if the goal is to assess the effect of a housing or tax policy on families, which aspects of family life should be, or can be, examined to determine the impact of the policy? What are the dependent variables for a case like housing policy? Should we measure its family impact by whether a person's marital status must be reported to establish eligibility for a federally insured mortgage? Or by whether the design specifications for public housing apartments include laundries and playgrounds? Or should altogether different considerations apply? Similarly, should we measure the family impact of tax policies by looking at whether one-earner and two-earner families are assessed at the same rates?

For this study of work schedules, in particular, further development of conceptual models could focus on refining both the dependent and independent variables and their interactions. That is, the efforts could be directed at refining the definitions of the three elements of family functioning we utilized, namely, stress, time, and equity; or in identifying additional family functions. Or attention could be directed at identifying other work issues that may affect family life in addition to schedules.

The results of our study revive a number of old questions about family functioning that require continued development of conceptual models to produce useful empirical data. For example, what constitutes the substance and process of good child rearing and parenting, day in and day out, in a culture where both parents of the majority of children work away from their children most days and weeks of their growing up? What kind of parenting matters at what ages? Day-to-day, throughout the year, how much time do fathers and mothers spend doing things that really matter for healthy child development? Despite the abundance of research on stages of child development, extraordinarily little is known about the ecology of parent-child interaction as suggested by the questions and worries of parents in this study.

Repeatedly, employed parents in our interviews said that although they worried about whether they spend enough time with their children, they did not know how much time and what kind of time with their children "really counted." Fathers especially—in the context of expressing regret about being too pre-occupied with work *or* other concerns, and therefore spending too little time with their children—reiterated that they did not know what an extra hour here or there with the children "means" in terms of their children's development. In contrast, they knew exactly what more time means at work: pay, promotions, prestige, peer esteem.

Concurrently, the mothers, unlike the fathers, tended to have much clearer ideas about exactly what an extra hour with the children here or there "means," both in terms of attending to home chores and to children's needs. The mothers not only tended to worry more than the fathers about whether they were neglecting their children, they also tended to do more about the worry, in the sense of sustaining more regular involvement in daily child rearing. The fact that mothers more often know what constitutes "quality" time with their children apparently reflects the fact that they typically spend more time with them and therefore know from their own experience "what matters."

Clearly, the above differences between mothers and fathers suggest an important set of research questions about the interactions between the sex of parents, sex roles, and parenting and work values and activities. Differences between mothers and fathers on a whole range of work-family attitudes and behaviors remained a steady theme throughout the interviews.

On the independent variable side, more conceptual definitions of male/female work issues are also needed, along with theory, to hypothesize the interactions with family functioning. For example, what effects do increased rates of women's employment have on men's work and family roles? Characteristically, women have less demanding and absorbing jobs, even when they have comparable education and training. This disparity is due less to discrimination, in the views of our interviewees, than to the fact that they *chose* less demanding jobs because of their greater involvement in—and responsibility for—their children on a day-to-day basis. What effects do such choices have on women's work lives? What effect does the imbalance between mothers' and fathers' involvement in child rearing have on both the children and the parents? Can they be identified and measured?

What kinds of job environments enhance good parenting and which do not? What are the salient factors? Which aspects of employment are more or less amenable to change? What results can be predicted about various options in work patterns and policies? At what points in a family system or a social system is a policy intervention effective? And to what ends?

Few of these kinds of questions can be answered or even framed with current theories. Ecological and systems theories have laid important ground rules from which to begin to respond to these questions, namely, the insistence that family research be conducted contextually in terms of multiple influences, as it happens, and not in artificial laboratories. But these models need to be expanded by more detailed and testable hypotheses of how the behavior of mothers, fathers, and children is affected by work factors.

As the Family Impact Seminar has emphasized, and as the above examples suggest, with each attempt to specify criteria for measuring the family impact of a public policy, a question of values is raised. Thus, a major challenge and difficulty of this approach is to resolve the question of values. To pursue one of the above examples: if the analysts' personal or organizational values dictate that non-married couples should not be eligible for federal housing subsidies, then they may choose to measure the family impact of a housing policy differently than if they believed the opposite.

In this study, the Family Impact Seminar essentially took as a given that all workers care about their family lives. But the evidence from the data suggests that this does not necessarily mean that employed people want to spend more time with their families—or that they know how to spend more time satisfactorily, especially with their children. Thus, for future family impact analyses on this topic, it is necessary to refine conceptual ideas of what constitutes family time well spent.

The challenge of coming up with operational definitions of values may be equally difficult. Looking at the housing policy example again, the measures of family impact could be any of the following: whether the buildings are high rise or low rise; whether stores and services are accessible; whether public transportation is available and economical; whether there are criteria for a mix of age, race, income level, and family size. Resolving what to measure will lead the analysts or researchers back to asking why they should measure a particular phenomenon, that is, back to value questions.

Some distinctions among the above generalizations should be added. So far the frameworks seem most useful for analyzing policies aimed directly at families or their members (e.g., the elderly, teenagers, or infants). They may be less helpful for analyzing the effects of policies not primarily intended to affect family life. There are also many possible levels of analysis for different purposes and policies. Conceptualization of family factors for analysis may vary in complexity, depending on whether a policy is aimed at family life (e.g., child health, teenage pregnancy, and care of the elderly) or designed for other purposes, and therefore, only inadvertantly affecting families (e.g., tax policies, employment policies, and transportation policies).

For example, although foster care programs are aimed most directly at the well-being of children, it is a relatively short step, conceptually, to pose questions about how policies that remove children temporarily from their homes may also affect the biological family. The theory and mechanics of establishing flexitime programs, on the other hand, do not readily suggest consideration of issues related to families. Therefore, injecting family variables into flexitime research creates a new focus, requiring the development of more elementary, or perhaps more fundamental, ideas of the ways in which the two arenas—family and work—may be connected. A framework for assessing the impacts of policies with "indirect" effects may require an expanded model of Figure 10.

Yet, from another point of view, the frameworks may also be seen as most valuable for raising questions about policies that have only indirect or unintentional effects on families, like transportation policies or federal reserve discount rates, precisely because these policies are seldom made or implemented with family needs or values in mind.

In addition to illustrating some of the limitations of the analytic framework, the present case study of work schedules also suggests further complications that may arise in trying to conduct family impact analyses of policies on a wider scale. One of these complications is the expense of the process in terms of money and time. At a cost of more than $70,000 (taking into account the start-up costs; developing familiarity with related fields; a share of the Family Impact Seminar expenses for office rental, secretarial staff, and so forth; and the cost of research associates and assistants, printing, computer time, and the like), our work-family study is clearly not a model project for most citizen or governmental groups to emulate. The range of advisors, the extensive review of the literature, the research design, and data collection may be inappropriately expensive and complex to apply to a large number of social programs or policies.

Furthermore, the process of developing useful concepts and measures for this exercise (from congressional testimony and related family research) was complex and lengthy, as well as expensive. We found no useful concepts in the research literature for identifying the aspects of family life that might be affected by work schedules; inevitably, therefore, we did not find any measures of family life that could be suitably used in such a study—especially none that were readily adaptable to a self-administered questionnaire survey.

Extrapolating from this experience with our attempt to examine one relatively small dimension of family life that is affected by one minor government policy (namely, rules regarding hours of work), it seems unlikely that conceptual definitions for aspects of family lives that may be affected by other policies are readily available either. By the same token, neither will there be useful measures of the family issues. Thus, to conduct family impact analyses on many policies may require even longer and more arduous efforts to conceptualize which aspects of family life should be looked at in terms of how they are affected by policies.

For example, the conceptualization of the control variables for this study did not attempt to take into account all the ecological factors that doubtless play a role in affecting people's levels of stress and family work. On the workplace side, for instance, the design did not attempt to assess degrees of work absorption or pressures; or variations in job responsibilities, office functions, atmosphere, or supervisor styles. Similarly, although the work setting is described in Chapter 6, the design does not really enable us to calibrate the effects of its subtle but pervasive impact. Nor does it weigh the value of other supports to family life, like child care arrangements, the accessibility of shopping, and other community services and social support networks. The design also does not consider the qualities of marital relationships or personality characteristics that influence levels of stress and kinds of behavior. Some personalities may function well when their time is structured by forces outside themselves; others may thrive with more leeway. Work-family conflicts may be secondary to individuals' needs for relative freedom or order. Thus, to understand the impact of one policy on family life ideally requires being able to control for a multitude of other influences.

In addition to these problems of defining the "impacts" to be assessed, and of developing measurement techniques, the amount of time available for family impact analysis also will affect its potential usefulness on a wider scale. As discussed earlier, a longitudinal design—with the same families studied before and after flexitime—would have offered more chance of assessing whether a new work schedule seemed to be a cause of any changes in family life.[1] A lengthier study of that sort, however, with comparable numbers of people (700), would have increased the cost of the process even more (as well as delaying findings for another year or more, depending on how longitudinal the study became).

Yet for our modest topic of work schedules, the legwork we have done in order to conduct the survey may enable other groups to take some shortcuts when examining the family impact of work policies. We developed conceptual definitions of certain aspects of family life that may be affected by work schedules (family stress and family work); we designed measures and established reliability and validity for their use; and we collected data from seven hundred federal employees that offer some empirical information on how work schedules may affect family processes. Other groups can use our concepts and instruments to assess the impact of work schedules in other places.

Even the concepts and measures painstakingly pieced together for this study, however, are crude in terms of capturing the complexity of the ways in which policies affect families. For example, although the dependent variable *stress* was carefully defined for this study to refer only to tensions arising at the points at which people's work and family roles connect or overlap (see Chapter 4), the definition still leaves conceptual uncertainties, two of which are the following.

Would schedule flexibility have different effects on the kinds of work-family stress arising from physically strenuous work, such as digging with a jack hammer or holding two jobs (like the

[1]For example, early evaluations of the first Head Start programs suggested that initial IQ differences among experimental and control group children were not sustained by the end of the second grade (Westinghouse Learning Corporation 1969). But other studies that evaluated children from structured pre-school programs on different achievement measures ten years after the "early intervention" indicated that the experimental group of children was significantly different on several measures (e.g., less likely to be left back a grade in school and less likely to be in special education classes) from the control group, which did not participate in a pre-school program (Lazar et al. 1977).

mother of three who spends seven daytime hours on her feet in a beauty parlor and five evening hours cleaning our office building)? Perhaps the special kind of stress that we tried to pinpoint by our measures is relatively small for federal employees in contrast to other kinds of workers.

In another vein, as noted in Chapter 4, the theories of chronic stress used in this research also do not yet enable us to account for positive aspects of stress—for example, the sense of purpose, order, productivity, and well-being that can result from the challenge of, commitment to, and absorption in one or more roles. As Marks (1977) suggests, when people play multiple roles—for example, active participation in both job and family—they may feel vitalized rather than depleted by stress. The concept of stress in our study is a relatively simple one that cannot account for the more subtle interactions between time, responsibilities, energy, and the positive and negative dimensions of stress.

For all the reasons sketched above, the idea of carrying out family impact analysis should be viewed with caution. Yet as the Seminar's Interim Report emphasized, there should be ways of conducting the process without formal research designs. More thought and experimentation are required to sort out what this recommendation may mean in practice: who (in terms of organizations and training) can do what kind of family impact analysis with what goals and what results? In this connection, the Family Impact Seminar received additional funding from the Administration for Children, Youth, and Families in the Department of Health and Human Services to support several field experiments of family impact analysis with several different kinds of state and local organizations and civic groups in different parts of the country.

Despite the difficulties inherent in family impact analysis, perhaps the most valuable result of the Family Impact Seminar's efforts will be precisely and simply to get family-related questions raised among groups and institutions that, heretofore, have not considered the impacts of their activities on families.

With the considerations about families that are outlined in the framework tables of the Family Impact Seminar, almost any form of research could be used to conduct family impact analyses, as the Seminar's several case studies and field projects have demonstrated. Even more important, the process may provoke and facilitate responses to such impacts. In raising the national consciousness about the effects of work and other policies on families, the Family Impact Seminar has begun a challenging process of devising ways to assess how the policies of the public and private sectors affect citizens' family lives.

Appendix A

An overview of the literature

Family Research

On the family side of the equation, both government policies and academic research have addressed family issues in many different ways. For example, in terms of public policy, the federal government now has 268 programs which have direct effects on families, ranging from taxation to transfer payments, to marriage, custody laws, and educational subsidies, to health insurance. An additional group of policies at other governmental levels may have indirect effects on families, such as public transportation, work policies, and school starting and leaving ages (Family Impact Seminar 1978a and 1978b).

In academic research, as Skolnick (1975) and Troll (1969) have pointed out, any study of the family encounters a "limits of competence" problem because no single scholarly discipline can give an adequate picture of the family. Not only do many disciplines address family issues (e.g., economics, genetics, physiology, archaeology, anthropology, sociology, psychology and history), but the interdisciplinary nature of the subject presents epistemological problems for a family researcher. The assumptions of one discipline do not necessarily transfer directly to another field. For example, in Skolnick's (1974, p. 23) illustration:

> some biologists and physically oriented anthropologists tend to analyze human affairs in terms of individual motives and instincts; for them society is a shadowy presence, serving mainly as the setting for biologically motivated individual action. Many sociologists and cultural anthropologists, in contrast, perceive the individual as an actor playing a role

> written by culture and society. . . . [Some] psychologists see man neither as a passive recipient of social pressures nor as a creature driven by powerful lusts, but as an information processor trying to make sense of his environment.

Family Research Historically

Among the social sciences, probably sociologists have contributed the largest portion of family research. In the introduction to their recent review of predominantly sociological theories of the family, Burr and his co-editors (1979) outline their understanding of the evolution of family research in the last century and a half. Following Christensen (1964), they identify two major traditions in the one hundred years up to 1950. For the first half of the period, they characterize family research as dominated by historical essays attempting deductively to identify the universal characteristics of family life in all cultures.

From this kind of *macrotheorizing*, family research shifted, they explain, to more empirical, inductive, problem-oriented approaches after the turn of the century. By 1950, seven hundred books and articles annually were being produced on family topics worldwide (87 percent of these in English). American research concentrated on microstudies of contemporary issues like mate selection, marriage and divorce, sex and fertility, and family interaction with kin and the community. In the 1940s, Parsons (1947) led a major shift away from this problem-solving sociology of the family and into broad, abstract theorizing about family structure based on differing male-female functions (Burr et al. 1979, pp. 3–15).

In Burr's assessment, the period since 1950 has been "an era of systematic theory building" in family research. Following Merton's admonition to codify family theories (1945), a number of efforts at identifying conceptual frameworks for family study were undertaken. For example, Goode (1959) looked at the family as a determinant, as opposed to an accommodating entity among social institutions. As part of the Minnesota Inventory of Marriage and Family Research, Hill and Hansen (1960) identified five frameworks through which most family research had been approached: institutional, structure-function, symbolic interactional, situational, and developmental. (They dropped two frameworks from an earlier list: learning theory and consumption economics.)

Spurred by Zetterberg (1965), Nye and Berardo (1966) went on to consider the value and limitations of each of the frameworks designated in the Minnesota Inventory. Next, Broderick (1971) recommended dropping the situational and institutional frameworks and adding systems theory. Following a series of exchanges about strategies for family scholars (e.g., Aldous 1970), Burr (1973) attempted a "theory-repairing program," using both deductive and inductive procedures and borrowing from contiguous disciplines and general theories to deduce family phenomena. His 1979 volume, edited with Hill, Nye, and Reiss, continues this effort by means of a multi-university project that attempts to summarize and evaluate the current status of theories about families (Burr et al. 1979, pp. 7–15).

Yet, as Hodgson and Lewis (1979) point out in their review of family research literature, only about one-third of the family articles from 1962 to 1976 were guided by theory. For those studies with theoretical underpinnings, the dominant frameworks have been symbolic, interactional, structure-functional, and developmental; more recently, systems and developmental frameworks, emphasizing the interdependence of family members, have begun to incorporate or supplant the other two (which tend to be limited to socialization and personality development).

While most family research has been done by psychologists or sociologists, Hodgson and Lewis point out that few of the articles have appeared in the general journals in these fields; instead, the majority of reports of studies have appeared in three specialized journals, namely, *The Journal of Marriage and the Family*; *The Family Coordinator*; and *Family Process*. Their review article reports on studies in these three journals only.

Recent Family Research: Sociology and Psychology

Shifting economic, demographic, and social circumstances in recent years have raised new family research and policy issues. For this study, three particular developments—increased female employment, longer life spans, and dropping birth rates—have introduced new questions in family research; for example, how are young children raised when parents are employed away from home? Despite "a veritable explosion" of family research activity since 1965 (nearly 13,000 studies), until the mid-1970s little of this research focused on the role of families in child development, or on the impact of societal institutions on families' abilities to perform that role (Newbrough 1978). Similarly, until recently social scientists have paid little attention to the ways in which work—its locations, its demands, timing, values, and rewards—affect the family lives of workers (Furstenberg 1974; Gardell and Nilsson 1974; Kanter 1977).

In the last decade the bulk of social science research that deals with the connections between work and family has consisted of sociological and/or psychological studies in relation to four major topics (most of the literature mentioned below is discussed in more detail in Chapter 4 in relation to particular variables in this study): female employment, housework, child rearing, and sex roles and the division of labor in marriage.

1. *Female employment.* Most of the research and theory on women's employment in relation to family pursues aspects of the "two roles" theme (i.e., home and work) delineated in the mid-fifties by Myrdal and Klein (1956), including, for example: Bailyn (1973); Bernard (1972, 1974); Poloma and Garland (1971); Pleck (1977a, 1977b); Safilios-Rothschild (1970); Schopp-Schilling (1977); and Vanek (1977). Dahlstrom and Liljestrom (1971) and Michel (1971) edited collections of essays on various aspects of the "two role" topic. Others have collected empirical data from small samples of women on the same theme: Fidell (1976); Herman and Gyllstrom (1977); Hood (1977); Hunt and Hunt (1977); Frank A. Johnson and Colleen L. Johnson (1976); Orden and Bradburn (1969); Powell and Resnikoff (1976).

2. *Housework.* Following Gilman's 1903 essay on the nature of housework, relatively little attention was given to the substance of what nonemployed women do all day until the time-budget research of the 1970s (Szalai 1972; Robinson 1977a, 1977b; Meissner et al. 1975; Walker and Woods 1976; and Vanek 1974, who re-examined data collected earlier by the government). Recent case studies also have included interviews with housewives to learn more about the substance of their activities and their feelings about how they spend their time (Arvey and Gross 1977; Berk and Berheide 1977; Ferree 1976; Komarovsky 1964; Lopata 1971; and Oakley 1976). Mainardi (1972) among others, has discussed the "sexual politics" of housework, that is, the trends toward glorifying or denigrating it in different eras.

3. *Child rearing.* Studies relating work and child rearing have most

often considered the effects of maternal employment on children, indicating that employment itself has few negative effects on children, as long as the mother wishes to be in the labor force (for example, Cook 1975; Cox 1974; Hetherington et al. 1977; Hoffman and Nye 1974; Maccoby 1958; also literature reviews by Howrigan 1973; Etaugh 1974; Hoffman 1974; and Clarke-Stewart 1977). Several recent studies also critique the limited roles fathers have played in child rearing, in contrast to mothers, and some of the negative consequences of this imbalance (for example, Biller 1976; Fein 1978; Feldman and Feldman 1975; Gronseth 1971; and Hetherington 1972).

4. *Sex roles and the division of labor in marriage.* Since the women's movement in the 1960s, research on this topic has moved away from the Parsonian assumptions about instrumental and affective work roles for men and women, and has shifted to concern with the relationships between labor market participation and family roles—in terms of power, resources, satisfaction, and functions. For example, Blood and Wolfe (1960) and Clark and his co-authors (1978) have considered the effects of employment on male family roles; and Bahr (1974), Hoffman (1960), Lein and her co-authors (1974), Presser (1977), and Nickols (1976) have looked at the effect of female employment on the division of labor in the home; Bueche and Morotz-Baden (1978) have examined decision making in two-earner families; Bailyn (1970), the Rapoports et al. (1974), Orden and Bradburn (1969), and Haavio-Mannila (1971) have examined male and female satisfactions in relation to work and family; Cherlin (1978) has assessed the effects of work on divorce.

In addition to the four major groupings of work and family research suggested above, a few psychological and sociological studies have attempted to understand the multiple effects of work on the emotional life of a small number of families. Through small-scale, in-depth studies, Rubin (1976), Piotrkowski (1977), and Lein and her fellow authors (1974) have explored these issues in blue-collar families. In two other small-scale studies, Holmstrom (1972) and the Rapoports and their co-authors (1976, 1977) have looked at multiple characteristics of dual-career professional families. Other anthropological and ethnographic studies have characterized daily life in whole communities and families in order to understand various aspects of family life: Henry (1965) the characteristics of pathogenic families; Stack (1974) the coping strategies of poor, urban black families; Parker (1975) the psychopathology of two generations of one family; Wylie (1958) the work, education and social lives of families in a small French village.

Other Family Research

Outside these fields, research in several other disciplines has touched on work and family issues in relation to questions of particular interest to the discipline involved. Beginning in the late 1960s, for example, economists studied the interrelationships between labor market economics and home economics in terms of female labor force participation and other aspects of family economic decision-making (e.g., Epstein 1970; Hayghe 1976; Hedges 1972; Johnson and Hayghe 1977; Kreps 1976; Lloyd 1975; Macke 1977; Michel 1971; Moore and Sawhill 1976; Owen 1975, 1977b; Sexton 1977; Ralph E. Smith 1977a, 1977b, 1977c).

Research in medicine and mental health has considered child and adult development, mentally and physically, in relation to family dynamics and social structures (e.g., Brim 1975; Lewis 1976; Guttentag and Salasin 1975; Haley 1976; Hartup 1978; Minuchin 1975; and Turner 1970). In the last decade, historians, too, have evolved new ways to look at families which stem, in part, from questions raised by the other disciplines (see Chapter 2).

Work Research

Although work may seem a subject more amenable to study by a single academic discipline than families, it, too, is a subject characterized by "limits of competence" problems. The topic of work has hardly been the province of economists alone; it is also a favorite of theologians, philosophers, sociologists, historians, anthropologists, and psychologists, among others. The epistemological challenges in defining work are at least as complicated as those in studying families. For example, the meaning of work has been addressed philosophically in Western culture from ancient Greece to Christian theologies (e.g., Plato, St. Augustine, Aquinas, Luther, and Calvin), and by political and economic theorists alike (e.g., Rousseau, Locke, Smith, Carlyle, Ruskin, and Marx).

In the twentieth century, work has been considered in a multitude of additional ways. For example, within the general focus on labor market supply and demand questions, economists have considered work incentives, unionization, education and training, compensation, and quality of employment—among other issues (e.g., Berg 1971; Lloyd

1975; Sawhill 1974; Ralph E. Smith 1977c). In terms of public policy, work questions are addressed in terms of human capital, regulatory and monetary policies, by social insurance and welfare programs, and taxes and grants (e.g., Levitan, Magnum, and Marshall 1972; Etzioni and Atkinson 1969). Sociologists have studied work in relation to social structures, social class, and leisure (e.g., Rainwater 1970; Gans 1962; Becker 1965; Best and Stern 1977; Caplow 1954; De Grazia 1962; Linder 1970; Blau and Duncan 1967; Clayre 1974). Psychologists have related work content, performance, settings, and organization to people's intellectual and emotional needs, to their aptitudes and satisfactions, and to their identities (e.g., Borow 1967; Crites 1969; Freud 1905/43; Hamburger and Hess 1970; Levinson 1964, 1971; Maslow 1954, 1962, 1970; Strong 1943; Super 1962). Bernard (1974, p. 112) has ranked the "nature of work," along with the "nature of sex" as the major human puzzle, noting that, "if it is baffling in the case of men, it is doubly so in the case of women." (See also Sokolwoska 1965, and both the citations on women's employment and the discussions of the dependent variables in this study).

Appendix B

Bronfenbrenner's ecological theory

One of the conceptual frameworks that seemed best suited to our investigation of work schedules' effects on families was the ecological approach to human development of Bronfenbrenner (1977b), which is derived from Lewin's topological theories (1935, 1936, 1948, 1951). This orientation to understanding family functioning assumes that human beings must be understood in the context of the relationships both in their immediate and also in wider social environments. The ecology of human development is defined as the scientific study of the progressive, mutual accommodation throughout the life span between a growing human organism and the changing environments in which it lives. This outlook sees families as one kind of institution, or unit, among many social institutions in the environment—such as schools, health facilities, neighborhoods, churches, stores, transportation, and workplaces. Bronfenbrenner defines the ecological environment as a nested arrangement of four structures, or systems, each contained within the next.

First is the *microsystem*, which includes the relations between the developing person (children and adults) and the environment. These relations take place in an immediate setting containing that person (e.g., home, school, workplace). A *setting* is defined as a place with particular physical features in which the participants engage in particular activities in particular roles (e.g., daughter, parent, teacher, employee) for particular periods of time. Elements of a setting include place, time, physical features, activity, participant, and role.

Second is the *mesosystem*, which comprises the interrelations among major settings containing the developing person at a particular point in his or her life. A mesosystem is a system of microsystems. For example, for an American twelve-year-old, the mesosystem typically encompasses interactions among family, school and peer group; for some children it might also include church, camp, or the workplace.

Third is the *exosystem*, which is an extension of the mesosystem, embracing other specific social structures which impinge on or encompass the immediate settings in which a person is found, and thereby influence, delimit, or determine what goes on there. These structures include the world of work, the neighborhood, the mass media, agencies of government, distribution of goods and services, communications and transportation facilities, and informal social networks.

Fourth is the *macrosystem*, which refers to the overarching institutional patterns of the culture such as economic, social, educational, legal and political systems, of which micro-, meso-, and exosystems are concrete manifestations. Macrosystems are conceived of and examined not only in structural terms, but also as carriers of information and ideology that give meaning and motivation to particular agencies, social networks, roles, activities and their interrelations. For example, the place and priority of children and those responsible for their care in such macrosystems is of special importance in determining how a child and his or her caretakers are treated and interact with each other in different types of settings.

Within this ecological framework, the present study focuses on the interactions between one microsystem, the family, and one exosystem, the world of work; together they are shaped by the overarching institutional patterns of the macrosystem in which they exist in this culture. In particular, our research examines one element of the workplace structure, the work schedule, in terms of its effects on two aspects of family life, namely, family work, and family stress.

Appendix C

Scales to measure stress in relation to work and family roles

The conceptualization of each scale is described separately, and the discussion of their construction and the establishment of reliability and validity are presented together. Various members of the Advisory Committee on the study helped with the development of these scales at various stages, including Mary Jo Bane, Urie Bronfenbrenner, John Demos, Jerome Kagan, Sheila B. Kamerman, Rosabeth Moss Kanter, Robert K. Leik, Joseph Pleck, Mary Sue Richardson, and Isabel V. Sawhill.

Job-Family Role Strain Scale

We developed a role strain scale in order to assess people's feelings as both family members and workers when what is expected of them in the two roles overlaps or conflicts. The terms *role strain* and *role conflict* have both been used in the literature to describe this phenomenon, but distinctions between the two terms are not consistent among different authors (e.g., Goode 1960; Rapoport and Rapoport 1965). For this study the following clarification seems most useful. Role conflict includes only inter-role and intra-role conflict (Sarbin and Allen 1968; Nye 1976); whereas role strain is a wider concept which includes conflict along with several other dimensions.

Although a number of studies deal with work-family role strains (Blood and Wolfe 1960; Burchard 1954; Gross, Mason, and McEarchern 1958; Hoffman and Nye 1974; Nevill and Damico 1974; Weller 1968; Wispé 1955), the measures of strain in these studies usually apply separately to men or to women; none attempts to assess role strain related to job and family responsibilities by a single measure for both males and females. Among the several hundred family measurement techniques described by Strauss (1969), several touch on the work-family role strain issues of interest to the present study, but like many marital satisfaction measures (e.g., Spanier, 1976), these instruments focus on interpersonal issues without addressing the tensions emerging from the structure of the roles.

For example, the eighty items in the Moos "Family Environment Scale" (1974) are designed to measure and describe the interpersonal relationships among family members, the basic organization of the family, and the directions of personal growth emphasized in each family. The Pless and Satterwhite "Family Functioning Index" (1973) is intended to assess the strength of relationships and life styles of whole families. The thirty-five items were designed for physicians to use in interviews with parents of ill children. The eighty-item VanderVeen "Family Concept Test" (n.d.) was also developed for medical use and consists of characteristics on which respondents can indicate similarity or dissimilarity in their own families.

Although Komarovsky's (1977) formulation of role strain was developed for one sex (male college seniors), it is a comprehensive conceptualization, broad enough to encompass role strains emerging from structural work and family demands on both men and women. Komarovsky defines role strain as people's feelings of discomfort or tension about whether their performance of specific roles lives up to their internalized expectations. She designates six modes in which such discomforts or tensions may occur (Komarovsky 1977, pp. 226–237). The following list of Komarovsky's modes is illustrated by examples from the present study, thus illustrating the applicability of her conception to the family work topic.

1. Ambiguity about which norms regulate a certain situation (for example, the allocation of domestic tasks).
2. Lack of congruity between an individual's personality and a particular social role (for example, when a housewife's need for achievement is frustrated by housewifery).
3. A socially structured insufficiency of resources for role fulfillment (for example, when a single parent cannot find suitable child care facilities).
4. Low reward for role conformity (for example, when a woman feels her home-based work carries less social esteem than employment).
5. Conflict between normative role phenomena (for example, if a fa-

ther must be at work when other family members are available to be with him).

6. Overload of role obligations (for example, when an individual has too many statuses—such as parent, student, child, friend, spouse, worker, and community leader—to meet the demands of each status to the satisfaction of all the role partners and the satisfaction of self).

Since no existing measures successfully combine the several issues to be assessed together in our study, the Job-Family Role Strain Scale was developed following five of Komarovsky's six modes (excluding number two because the scope and design of the study could not adequately take account of personality variables). The items for each mode appear below. They include one group which can be answered by parents and non-parents, and a second group about child care for parents only:

Mode 1. Ambiguity about norms:

1. I worry that other people at work think my family interferes with my job.
2. I worry whether I should work less and spend more time with my children.
3. I worry that other people feel I should spend more time with my children.

Mode 3. Socially structured insufficiency of resources for role fulfillment:

4. I worry about how my kids are while I'm working.
5. I am comfortable with the arrangements for my children while I am working.
6. Making arrangements for my children while I work involves a lot of effort.

Mode 4. Low rewards for role conformity:

7. I feel more respected than I would if I didn't have a job.
8. I am a better parent because I am not with my children all day.
9. I don't have as much patience with my children as I would like.

Mode 5. Conflict between normative phenomena:

10. My job keeps me away from my family too much.
11. I have a good balance between my job and my family time.
12. My time off from work does not match other family members' schedules well.
13. I always find enough time for the children.

Mode 6. Overload of role obligations:

14. I feel I have more to do than I can handle comfortably.
15. I wish I had more time to do things for the family.
16. I feel physically drained when I get home from work.
17. I feel I have to rush to get everything done each day.
18. I feel I don't have enough time for myself.
19. I feel emotionally drained when I get home from work.

Thus, the Role Strain Scale seeks to tap the kind of stress related to internalized values and emotions—such as self-doubt, worry, guilt, and pressure—but also feelings of contentment, fulfillment, self-respect, and balance in regard to job and family obligations. The scale enables employed persons to indicate (on a Likert format) how often they feel strains of various kinds related to time for job and time for family. Higher sum scores indicate greater job-family role strain. (See the order of items and format in the instrument in Appendix H, questions 20 and 41.)

Family Management Scale

The Family Management Scale is concerned with those routine and special activities—appointments, events, errands, visits, and family interactions—that employed persons must manage outside their hours of work. These activities may involve interactions with schools, health services, service organizations, stores, libraries, or various other institutions or individuals in the social environment in which the family functions. On days when a person is working on a job, his or her ability to interact with or on behalf of other family members will depend in part on the work schedule that defines when the person may or may not be present on the job.

The ways in which work schedules affect the family management issues characterized above have been suggested in several recent publications. In her overview of the work-family interfaces as a research field, Kanter (1977) has pointed out that family events and routines are built around work rhythms, just as much as the timing of events in the society as a whole (e.g., store and television hours) are predicated on assumptions about the hours, days, and months when people are most likely to be working or not working. The complexities faced by contemporary families with respect to the need and desire of members to interact with many other social institutions—for example, for Social Security payments, for school programs, for health insurance—have been suggested in recent publications by Keniston and the Carnegie Council on Children (1977), the National Academy of Sciences (1976), and the Family Impact Seminar (1978a and b).

Research methodology on these family management issues remains at an early stage of development. For example, recent exploratory studies have utilized open-ended interview questions on work effects on families (e.g., Bronfenbrenner's ongoing study at Cornell). Two Michigan Survey Research Center studies, the Quality of Employment Survey and Men's Roles Survey (Survey Research 1977a and 1977b), have included single-item questions related to the topic. Also another study with government volunteers used time logs to collect longitudinal data on total use of family time (Winett 1978). Another current inquiry uses a large number of single items to ask about the effects of flexible and shortened work weeks on both workplace and family factors (Cohen et al., ongoing John Hancock study).

But until now, no systematic scales have been developed on these family management issues, as far as can be determined from our wide-ranging review of the literature and conversations with scholars in the major centers of family research in the United States. To advance the methodology in this area, the Family Management Scale asks respondents to indicate how difficult or easy (on a Likert format) it is for them to perform or take part in a number of specific family and individual activities that occur outside their hours of work.

In contrast to the Role Strain Scale, which taps respondents' internalized values and feelings of self-doubt, guilt, worry, or pressure in regard to certain family and work issues, the Management Scale taps respondents' feelings about the logistics of family life—how easy or difficult, how simple or complex it is to accomplish certain activities.

The items for this scale have been developed for employed persons with or without children to obtain information on family management experiences from various family structures. Most of the items fall into several substantive clusters: health, education, retail services, commuting, and family and community interaction, as shown in the list below. For each item respondents indicate on a scale of 1–5 how difficult it is for them to have time for the particular activity. Higher sum scores indicate greater difficulty in family management. This scale also has a group of questions for all respondents and additional items for parents only:

Health:

1. To go to health care appointments
2. To take your children to health care appointments
3. To stay at home with a sick child

Education/child care:

4. To take your children to or from a child care setting or school
5. To go to school events for your children
6. To make alternate child care arrangements when necessary
7. To make arrangements for children during summer vacation

Retail services:

8. To go on errands (e.g., shoe repair, post office, car service)
9. To go shopping (e.g., groceries, clothes, drug store)
10. To be home for services or deliveries (e.g., telephone, appliances)

Commuting:

11. To avoid the rush hour

Family interaction:

12. To adjust your work hours to the needs of other family members
13. To have meals with the family
14. To spend fun or educational time with the family
15. To have relaxed, pleasant times with your children
16. To be home when your children get home from school

Community interaction:

17. To visit or help neighbors or friends
18. To participate in community activities

General overlapping items:

19. To go to work a little later than usual if you need to
20. To make telephone calls for appointments or services
21. To take care of household chores

Construction of Scales

Items for the two scales were generated by the following three methods. First, family members' statements from five studies were reviewed and sorted in terms of areas in which respondents reported strain in performing family and work roles and perceived difficulties in managing family activities. The statements came from the following sources:

- An exploratory study of the effects of flexitime on family life (Winett 1978).
- A pilot study of stresses and supports experienced by sixty-six fam-

ilies with young children in New York State (Bronfenbrenner et al. 1977).

- A study of environmental influences on child rearing practices of seventy-two divorced mothers at two income levels (Colletta 1978).
- A study of factors affecting job satisfaction for fifty-one women who had entered the labor force in the last two years after a period of time at home rearing children (Coiner 1978).
- A study of the origins of social stress with twenty-three hundred Chicago residents (Pearlin 1975, 1977, 1978).

Second, individual and group conversations of varying lengths (ten minutes to one hour) were held with fathers, mothers and children from ten families. Discussants were asked to describe the kinds of strains they felt in trying to be both good workers and good parents (or in the cases of children, they were asked to describe the kinds of strains they thought their parents felt).

Third, statements developed from the above two procedures were shown in written form to two groups of federal employees (six persons in each) in two federal agencies. These federal workers discussed whether these statements reflected their own feelings and any different feelings they had. As the items were generated, they were divided into the Role Strain or the Management Scale depending on whether they seemed to concern internalized values and feelings or feelings about the logistics of family life.

Validity and Reliability

The initial content validity of the two scales was established by a review of the items by a panel of six judges (two psychologists, a sociologist, and three federal personnel experts—Thomas Cowley, Barbara Fiss, Rosabeth Moss Kanter, Virginia Hilder Martin, Joseph Pleck, Mary Sue Richardson) who rated the items according to how well they tapped the content designated for the scale. Items which were approved by this process were included in the scales for pre-testing.

Pre-test. Reliability of the two scales was established in a pre-test on a sample of fifty federal employees from three agencies in Washington, D.C. The alpha coefficients were .71 for the Role Strain Scale and .93 for the Family Management Scale.

Concurrent validity was established by correlating respondents' scores on each scale with their score on a set of predictor variables.

Positive relationships were found between the degree of role strain and the number of hours worked by the respondent, the length of time spent commuting, and the number of hours worked by the respondent's spouse. Similar results were found for the relationships with the Family Management Scale.

Final Form. Following suggestions received during the pre-test, a few items were modified in each scale. For example, in the Job-Family Role Strain Scale, the item concerning physical and emotional fatigue was separated into two questions; in the Job-Family Management Scale the item regarding ease of being home for service calls was eliminated (see Appendix I, Figure 2).

During the review and analysis of the responses to the items on the final version of the Family Management Scale, it was concluded that the scale still had measurement problems, primarily in terms of respondents' interpretation of the question "how easy or difficult is it for you to arrange your time to do each of the following." The problem is that it is unclear whether the scale measures the objective difficulty people think they have in managing the designated responsibilities, or whether it measures a person's subjective reaction to simply having those responsibilities.

For example, a man who has not typically done most of the activities on the list may assume that such responsibilities would be very difficult for him, and his report that it would be difficult to do any of them may reflect the fact that it seldom has occurred to him that he should or could do them (e.g., grocery shopping, household chores). Women, on the other hand, may answer more "objectively" than men in the sense that they more typically assume responsibility for family activities and therefore can more realistically assess how easy or difficult it is for them to do these things on days when they are working. Given the very high inter-item reliability on this scale in the pre-test with fifty federal employees ($r = .91$), this interpretive problem did not appear until collection of the much larger sample size in the study itself.

Each scale was divided into two parts. The "Adult" part can be answered by respondents with or without children (questions 19 and 20 in Appendix H). The items in the "Parent" part are only relevant for people with children. The "Total" scales include the "Adult" and "Parent" items (questions 41 and 42 in Appendix H).

Reliability. The reliability for the Family Management Scale in the final form used in the study was still high for both versions of the scale.

TABLE 2
RELIABILITY COEFFICIENTS FOR THE STRESS SCALES*

Sample	Family Management Scale				Job-Family Role Strain Scale			
	Adult Scale		*Total Scale*		*Adult Scale*		*Total Scale*	
	α	N	α	N	α	N	α	N
Total sample	.884	(449)	.915	(92)	.723	(481)†	.603	(212)†
Males	.889	(239)	.913	(52)	.640	(263)	.528	(113)
Females	.879	(208)	.923	(40)	.671	(170)	.549	(66)
Standard time	.841	(192)	.921	(51)	.632	(192)	.497	(79)
Flexitime	.901	(156)	.907	(41)	.662	(241)	.561	(100)
Males, standard time	.842	(101)	.904	(26)	.657	(115)	.437	(48)
Females, standard time	845	(91)	.932	(25)	.602	(77)	.552	(31)
Males, flexitime	911	(138)	.912	(26)	.626	(148)	.546	(65)
Females, flexitime	.886	(117)	.906	(15)	.714	(93)	.563	(35)

* Reliability coefficient (α) for total sample and sub-samples.
† Reliability scores for this scale were increased after eliminating three items with the lowest item to total correlation. The original reliability coefficients with the three items were: Adult Scale α = .649, N = 433, Total Scale α = .551, N = 180.

The alpha coefficients were .88 for the Adult scale and .91 for the Total scale. The Job-Family Role Strain Scale scores were slightly lower than in the pre-test (Adult = .65; Total = .55). To increase the reliability for the Role Strain Scale, three items with low content validity were eliminated in the analysis of the data and those which had the lowest item to scale correlation. All three items were in mode 4. Item number 7 seemed to tap a different dimension of work concern than those related to family more directly. Items number 8 and 9 were eliminated because the negative statements seemed to create confusion for some respondents. With the omissions, reliability coefficients for Role Strain Scale increased to .72 for the Adult scale and to .71 for the Total scale. (See Table 2.)

Concurrent validity. Concurrent validity for the scales in final form showed a similar picture for that obtained in the pre-test. For the Job-Family Role Strain Scale, higher degrees of strain were found among women and younger respondents; among those who reported spending more time in home chores; among those who work and commute more hours; among those who report greater work-family interference (r = .49); and among those who feel they have major responsibility for home chores and child rearing.

For the Family Management Scale, those who reported greater difficulty included those who work longer hours; who have more children; whose children are age six to eighteen; who work and commute longer hours; and who report greater work-family interference (see Tables 3 and 4).

Criterion validity. Since there appeared to be no other measure that tapped the same or sufficiently similar content to the scales, criterion validity could not be established by comparing them with other instruments. In particular, a social desirability measure, which may seem to offer useful comparisons, is inappropriate for these scales for the following reasons. Our scales deliberately address feelings about the intersections between two life areas which have traditionally been studied separately, namely families and work. The social desirability of behaviors in one realm may be, indeed often is, very different from or at odds with the social desirability in the other. For example, what may be socially desirable from the standpoint of family well-being may be the opposite of that which may be socially desirable in terms of doing well on the job. In fact, the intention of these scales is to examine precisely the points of tension between two separate and often contradictory sets of socially desirable behaviors. Therefore, a single social desirability scale of the type that has been developed to indicate attribution and denial of negative and positive traits in individuals (e.g., Crowne and Marlowe 1964) would not help to indicate the criterion validity of these scales.

Construct validity. Some methodologists (for example, Kerlinger 1973, p. 468) consider factor analysis to be one of the most powerful methods of construct validation. Cronbach and Meehl (1955) have discussed the difficulties in using factor analysis for construct validity. Others argue that different methods, for example, multifactor–multimethod matrix, are more helpful (e.g., Paradise 1980). The idea of factor analysis to establish construct validity is that it reduces a large number of measures (like the items in the stress scales for this study) to a smaller number of factors by discovering which ones "go together," or measure the same thing. For example, if the scale items in this study divide into factors that coincide with the elements of the construct they are intended to measure, then the factor analysis helps demonstrate that the measure has effectively tapped the content it was designed to identify.

To use factor analysis to establish construct validity for the two stress

TABLE 3
CONCURRENT VALIDITY FOR JOB-FAMILY ROLE STRAIN SCALE CORRELATION COEFFICIENTS AMONG SCALE AND PREDICTOR VARIABLES

Predictor Variable	*Adult Scale* Pearson *r*	*Adult Scale* Significance Level	*Total Scale* Pearson *r*	*Total Scale* Significance Level
Sex (VAR006)	.16 (547)	.0001	.20 (273)	.0001
Number of children under 18 living at home (SCL018)	.11 (563)	.004	.01 (269)	.422
Spouse works (VAR072)	.09 (838)	.035	.17 (221)	.007
Outside help (VAR089)	−.06 (430)	.113	−.032 (267)	.3
Age of youngest child (VAR010)	−.05 (540)	.128	−.09 (271)	.069
Family life cycle stage (SCL017)	.10 (574)	.009	−.09 (273)	.073
Respondent has main responsibility in home chores (VAR087)	.15 (426)	.001	.13 (266)	.015
Respondent's main responsibility with children (VAR102)	.15 (267)	.006	.17 (266)	.003
Respondent's weekly hours in home chores (SCL007)	.21 (206)	.0001	.18 (254)	.002
Respondent's percent time in home chores (SCL011)	.24 (224)	.0001	.23 (128)	.005
Respondent's weekly hours with children (SCL009)	.16 (257)	.005	.09 (256)	.07
Respondent's percent time with children (SCL012)	.23 (120)	.008	.24 (120)	.004
Perception of family-work interference (VAR127)	.49 (549)	.0001	.52 (271)	.0001
Age of respondent (VAR007)	−.11 (570)	.005	−.22 (271)	.0001
Number of hours worked (VAR027)	.08 (567)	.033	.07 (273)	.123
Number of hours working and commuting (TIM001)	.16 (550)	.0001	.11 (268)	.039
Number of hours at job and in family work (TIM002)	.24 (243)	.0001	.18 (242)	.002

TABLE 4

CONCURRENT VALIDITY FOR FAMILY MANAGEMENT SCALE CORRELATION COEFFICIENTS AMONG SCALE AND PREDICTOR VARIABLES

	Adult Scale		*Total Scale*	
Predictor Variable	*Pearson r*	*Significance Level*	*Pearson r*	*Significance Level*
Sex (VAR006)	−.01	.378	−.08	.126
Number of children under 18 living at home (SCL018)	.15 (542)	.0001	.28 (219)	.0001
Spouse works (VAR072)	.02 (352)	.357	−.09 (173)	.109
Outside help (VAR089)	−.01 (397)	.407	.83 (217)	.112
Age of youngest child (VAR010)	.1 (550)	.008	.07 (221)	.135
Family life cycle stage (SCL017)	.1 (553)	.011	.14 (222)	.016
Respondent has main responsibility in home chores (VAR087)	−.05 (394)	.157	.06 (219)	.192
Respondent's main responsibility with children (VAR102)	.02 (252)	.365	.1 (217)	.073
Respondent's weekly hours in home chores (SCL007)	.1 (372)	.028	−.03 (206)	.337
Respondent's percent time in home chores (SCL011)	.11 (210)	.054	.1 (110)	.141
Respondent's weekly hours with children (SCL009)	−.13 (242)	.022	−.2 (208)	.002
Respondent's percent time with children (SCL012)	.08 (113)	.208	.1 (104)	.160
Perception of family-work interference (VAR127)	.41 (528)	.0001	.42 (220)	.0001
Age of respondent (VAR007)	.14 (549)	.0001	.16 (220)	.008
Number of hours worked (VAR027)	.18 (544)	.0001	.18 (222)	.004
Number of hours working and commuting (TIM001)	.24 (527)	.0001	.23 (219)	.0001
Number of hours at job and in family work (TIM002)	.01 (228)	.495	−.08 (195)	.141

scales, several a priori expectations were set forth about how the scale items would divide into factors, using a principal components analysis program with varimax rotation. Each of two stress scales was considered in three parts, one for all adults, one for parents only, and a total scale that combines the first two.

The most general expectation was that all the clusters of factors for the six versions of the scales would reflect the kind of stress being investigated in this study, namely, the experiences of discomfort, pressure, tension or frustration that arise as people function in both their job and family worlds. (See discussion of the Bronfenbrenner and Pearlin theories of stress utilized in the conceptualization of the study in Chapter 4.) This expectation was borne out in the factoring, and is explained for each version of the scale in the following discussion.

Factoring the Family Management Scale

For the Adult Family Management Scale it was expected that the items would factor into three major clusters: (1) concerning the logistical aspects of those activities that have to be done by most people (e.g., errands or health appointments); (2) relating to more optional activities which people may choose to do if their schedules permit (e.g., seeing friends or participating in community activities); and (3) having fun or educational time with the family.

These predictions were fulfilled in the factor analysis. The first factor, explaining 71 percent of the variance, consisted largely of items related to necessary activities (e.g., health appointments and errands). The second factor, explaining 18 percent of the variance, consisted largely of more optional activities (seeing friends, participating in the community, and doing chores). The third factor, explaining 11 percent of the variance, consisted almost entirely of having fun or educational time with the family. Thus, the construct validity of this scale seems substantiated by the factor analysis. (See Table 5.)

Similarly, the Parent Family Management Scale factored as predicted, that is, into one factor in which all of the items are related to aspects of child rearing, including direct care, appointments and relaxation. (See Table 6.)

The Total Family Management Scale was expected to factor into four clusters, reflecting a combination of the adult and parent versions. In fact, five factors appeared in the varimax rotation. Sixty-four percent of

TABLE 5
FACTOR ANALYSIS OF ADULT FAMILY MANAGEMENT SCALE

Characteristics	*Factor 1*	*Factor 2*	*Factor 3*
Eigen value	4.70	1.20	0.71
Percent of variance	71.10	18.20	10.00
Cumulative percent	71.10	89.20	100.00
Varimax rotated factor matrix variable			
VAR044: Avoid rush hour	0.33	0.00	0.07
VAR045: Arrive at work late	0.67	−0.01	0.17
VAR046: Health care appointments	0.66	0.28	0.16
VAR047: Errands	0.71	0.42	0.11
VAR048: Shopping	0.55	0.51	0.14
VAR049: Telephone calls	0.47	0.30	−0.22
VAR050: Household chores	0.13	0.69	0.23
VAR051: Visit friends	0.17	0.79	0.29
VAR052: Community activities	0.14	0.68	0.51
VAR053: Adjust work hours to family	0.57	0.18	0.45
VAR054: Eat with family	0.10	0.30	0.57
VAR056: Education fun with family	0.22	0.26	0.84

TABLE 6
FACTOR ANALYSIS OF PARENT FAMILY MANAGEMENT SCALE

Characteristics	*Factor 1*
Eigen value	3.95
Percent of variance	100.00
Cumulative percent	100.00
Factor matrix variable	
VAR118: Child health care appointments	0.78
VAR119: Getting to school	0.73
VAR120: Going places with children	0.72
VAR121: Child school events and appointments	0.81
VAR122: Alternative child care	0.69
VAR123: Being home after school	0.28
VAR124: Being with sick child	0.68
VAR125: Plan summer vacations	0.65
VAR126: Relaxing with children	0.41

the variance appeared in the first factor, and the dominant items in this factor concerned a variety of necessary child care activities (e.g., getting children to school, managing health care and school related appointments, caring for sick children, and arranging summer school vacation care). The second factor, explaining 14 percent of the variance, contained primarily items having to do with the fun and relaxing aspects of family life. Factor three, with 10 percent of the variance, was related to chores and activities with friends and community. Factor four (7 percent of the variance) consisted mostly of adjusting work hours to family along with some of the child care logistical issues. Factor five, explaining 5 percent of the variance, concerned the logistical aspects of health care appointments, errands, shopping, and telephone calls. (See Table 7.)

Thus, although the factors for the total scale varied somewhat from those predicted, the items still clustered in generally expected categories, with child care requirements explaining the largest part of the variance. The results of the factor analysis are in accord with the general theories and hypotheses for the study: first, that parents with direct child care responsibilities would feel significant amounts of stress related to balancing their family and job responsibilities; and second, that family events and routines are built around work rhythms, and people's ease or difficulty in interacting with or on behalf of other family members depends, in part, on work schedules which define when the person may or may not be present on the job. (See Kanter 1977; Keniston and the Carnegie Council on Children 1977; the National Academy of Sciences 1976; and the Family Impact Seminar 1978b).

Factoring the Job-Family Role Strain Scale

In addition to using the Bronfenbrenner and Pearlin theories of stress, the Role Strain Scale was conceptualized in terms of the six role strain modes identified by Komarovsky (1977) (see pp. 231–232). To establish construct validity for this scale, it was predicted that the scale items would factor into approximately the six Komarovsky modes. In fact, in the varimax rotation, the items did not factor perfectly into these six modes; but the three versions of the scale do have factorial clusters which coincide with five of the six modes.

In the Adult Role Strain Scale, 87 percent of the variance was accounted for by one factor, namely, Komarovsky's mode 6, overload of role obligations; in particular, the two items asking people if they felt

TABLE 7
FACTOR ANALYSIS OF TOTAL FAMILY MANAGEMENT SCALE

Characteristics	*Factor 1*	*Factor 2*	*Factor 3*	*Factor 4*	*Factor 5*
Eigen value	7.76	1.62	1.20	0.82	0.62
Percent of variance	64.50	13.50	10.00	6.80	5.20
Cumulative percent	64.50	78.00	88.00	94.80	100.00
Varimax rotated factor matrix variable					
VAR044: Avoid rush hour	0.12	−0.08	0.05	0.43	0.01
VAR045: Arrive at work late	0.12	0.09	−0.01	0.57	0.28
VAR046: Health care appointments	0.19	0.16	0.18	0.48	0.45
VAR047: Errands	0.25	0.23	0.17	0.49	0.58
VAR048: Shopping	0.21	0.26	0.30	0.25	0.62
VAR049: Telephone calls	0.21	−0.09	0.09	0.06	0.69
VAR050: Household chores	0.15	0.24	0.64	0.10	0.20
VAR051: Visit friends	0.31	0.19	0.78	0.18	0.15
VAR052: Community activities	0.26	0.43	0.68	0.16	0.12
VAR053: Adjust work hours to family	0.11	0.32	0.26	0.67	0.13
VAR054: Eat with family	0.20	0.72	0.15	0.04	0.10
VAR056: Education fun with family	0.05	0.79	0.28	0.28	0.06
VAR118: Child health care appointments	0.68	0.12	0.07	0.45	0.23
VAR119: Getting to school	0.77	0.04	−0.01	0.32	0.13
VAR120: Going places with children	0.57	0.17	0.32	0.31	0.08
VAR121: Child school events and appointments	0.60	0.21	0.40	0.22	0.24
VAR122: Alternative child care	0.63	0.16	0.34	−0.05	0.17
VAR123: Being home after school	0.13	0.24	0.17	0.22	−0.10
VAR124: Being with sick child	0.53	0.23	0.13	0.24	0.20
VAR125: Plan summer vacations	0.58	0.27	0.26	−0.05	0.12
VAR126: Relaxing with children	0.25	0.62	0.17	−0.06	0.07

TABLE 8
FACTOR ANALYSIS OF ADULT ROLE STRAIN SCALE*

Characteristics	*Factor 1*	*Factor 2*
Eigen value	3.47	0.50
Percent of variance	87.30	12.70
Cumulative percent	87.30	100.00
Varimax rotated factor matrix variable		
VAR057: Job keeps from family	0.31	0.64
VAR058: More than can do comfortably	0.49	0.40
VAR059: Good family-work balance	−0.25	−0.61
VAR060: Want more family time	0.29	0.55
VAR061: Work physically draining	0.85	0.26
VAR062: Work emotionally draining	0.71	0.23
VAR063: Must rush to get things done	0.47	0.41
VAR064: Time off doesn't match others	0.11	0.41
VAR065: Not enough time for self	0.35	0.40
VAR066: Others' view of family/job interference	0.08	0.20

*Reduced

physically and emotionally drained. The remainder of the variance was explained by a second factor, coinciding primarily with Komarovsky's mode 5, that is, conflicts with normative phenomena (e.g., if the job keeps a person away from family, if job and family seem well-balanced, if the respondent wants more time with the family, or if the respondent's time off does not match the family's time; see Table 8).

For the Parent Role Strain Scale there was only one factor that explained 100 percent of the variance. The items with the highest factor loadings reflected three of Komarovsky's six modes: (1) ambiguity about norms (mode 1: e.g., respondents worrying if spending enough time with kids, worrying about children while working, or worrying about others' views of how much time respondent spends with the children); (2) socially structured insufficiency of resources for role fulfillment (mode 3: e.g., respondents arranging for child care is difficult, or not being comfortable with child care arrangements); and (3) conflict between normative phenomena (mode 5: e.g., respondents finding or not finding enough time for the children; see Table 9).

For the Total Role Strain Scale, four factors were identified in the varimax rotation. Sixty-six percent of the variance appeared in the first factor and the relevant items were all in Komarovsky's overload mode 6 (the physically and emotionally draining items). For the second factor (18 percent of the variance), the items were mainly in Komarovsky's mode 5, conflict between normative phenomena (difficulty balancing job and family, or job keeping respondent from family); and mode 1, ambiguity about norms (e.g., worrying about the children, and finding time for them). In general, this second factor can be characterized as reflecting people's wish to have more time with their families. Factor three (9 percent of the variance) emphasizes mainly worry and logistical problems related to child care. Finally, factor four (7 percent of the variance) picks up several additional overload issues (feeling rushed, and having too much to do comfortably; see Table 10).

Clearly, the factor analysis of the three reduced versions of the Role Strain Scale supports the construct validity of the scale as reflecting discomforts, tensions, and so forth about balancing job and family roles successfully—in terms of Bronfenbrenner and Pearlin's chronic stress theories (see Chapter 4). Similarly, five of the six modes of Komarovsky on which the scale items were developed appeared in the factor analysis.

Moreover, in the complete versions of the Total and Adult scales, the fourth mode, low reward for role conformity, also appeared (e.g., with

TABLE 9
FACTOR ANALYSIS OF PARENT ROLE STRAIN SCALE*

Characteristics	*Factor 1*
Eigen value	1.54
Percent of variance	100.00
Cumulative percent	100.00
Factor matrix variable	
VAR110: I worry if enough time with kids	0.70
VAR112: I find enough time for children	−0.35
VAR113: I worry about children while I work	0.61
VAR115: I am comfortable with child care	−0.35
VAR116: Arranging for children difficult	0.46
VAR117: I worry about others' views of my time with child	0.47

*Reduced

TABLE 10
FACTOR ANALYSIS OF TOTAL ROLE STRAIN SCALE*

Characteristics	*Factor 1*	*Factor 2*	*Factor 3*	*Factor 4*
Eigen value	4.34	1.16	0.56	0.48
Percent of variance	66.30	17.70	8.60	7.30
Cumulative percent	66.30	84.10	92.70	100.00
Varimax rotated factor matrix variable				
VAR057: Job keeps from family	0.39	0.62	0.09	0.03
VAR058: More than can do comfortably	0.49	0.28	−0.04	0.44
VAR059: Good family work balance	−0.23	−0.66	−0.10	−0.23
VAR060: Want more family time	0.36	0.44	0.34	−0.01
VAR061: Work physically draining	0.82	0.20	0.18	0.10
VAR062: Work emotionally draining	0.72	0.20	0.05	0.06
VAR063: Must rush to get things done	0.46	0.21	0.11	0.50
VAR064: Time off doesn't match others	0.08	0.32	0.20	0.27
VAR065: Not enough time for self	0.36	0.21	0.35	0.21
VAR066: Others' view of family/job interference	0.05	0.06	0.26	0.18
VAR110: I worry if enough time with kids	0.15	0.44	0.61	−0.05
VAR112: I find enough time for children	−0.10	−0.48	−0.15	−0.18
VAR113: I worry about children while I work	0.03	0.11	0.60	−0.05
VAR115: I am comfortable with child care	−0.17	−0.11	−0.29	−0.32
VAR116: Arranging for children difficult	0.08	0.10	0.46	0.20
VAR117: I worry about others' views of my time with child	0.07	0.02	0.50	0.10

*Reduced.

the item, "I feel more respected than I would if I didn't have a job"). This item was dropped to increase the reliability of the scale; it did not appear to work well largely because it seemed to apply only to women who felt they had a choice about working as opposed to most male respondents who could not imagine not having jobs and therefore basically did not understand the question.

Appendix D

Measurement of family work and equity in family work

Family Work

While numerous researchers have studied stress and child development in a wide variety of ways (e.g., in laboratories and schools), relatively little scholarly research has examined the substance of family work until recently. From the turn of the century until the 1960s, the efforts were spotty: Gilman's classic statement, *The Home: Its Work and Influence* (1903), pointed out the interconnections between child care and house care; Bernard (1949), Lopata (1971), and Komarovsky (1964) touched on housework as part of larger sociological studies; the Bureau of Home Economics conducted a series of studies in the 1920s on how rural women budgeted their time (and these guidelines were used for more studies in the next several decades, e.g., Glazer-Malbin 1976; Vanek 1974; Myrdal and Klein 1956).

Assessing the amount of time employed persons spend on the job has been relatively easy in the twentieth century. Labor unions and management have monitored working hours. Mass standards, controls and output measures provide both employers and researchers with generalized indicators of what goes on and how long it takes. Calculating the time spent in family work, on the other hand, has been difficult for several reasons that have helped to conceal the total amounts of time and energy spent in domestic activities.

First among them is the fact that housework and child care in contemporary industrial countries tend to be carried out privately, in physical isolation, and are, therefore, self-defined. Although friends, neighbors and relatives are important sources of standards, family work is done in individual homes, and norms are divergent even in small communities. For example, as Oakley suggests, "meals can be cooked or cold; clothes can be washed when they have been worn a few hours or a few weeks; the home can be cleaned once a month or twice a day" (Oakley 1976, p. 8; see also Berheide and Berk 1977; Berheide and Berk 1976).

Second, family-related activities are so densely integrated and intertwined—chores, child rearing, marital relations, and the like—that separation of functions for study is never pure. Mothers mopping the floor or cutting up vegetables may be also listening for a baby's cry or discussing homework with a school child; indeed, most family work probably involves simultaneous multiple activities (see, for example, Bernard 1974, chap. 7; Lopata 1971). Social scientists may have been less successful at conveying the complex simultaneity of these activities than various women novelists and humorists (e.g., Drabble 1971, 1976; Roiphe 1968; Kerr 1960; Bombeck 1978; Viorst 1968).

Finally, a set of psychological factors has helped obscure the facts about home chores. Attitudes toward housework exist in the context of the larger universe of paid work in the society. As more women enter the labor force, housework appears to be increasingly devalued, and people are more ambivalently attached to it. High anxiety over low status work results in distortion of reporting on it. Berheide and Berk (1977) found that while women in their study reported hating housework, or denying that they did any, they simultaneously expressed satisfaction that research on it would inform others—particularly their husbands—about how much time and energy it took. Furthermore, because of the prevailing values about child rearing outlined above, child rearing is often given as the important reason for doing housework, reflecting the assumed social desirability of investment of time and energy in children, as opposed to housework.

In the last twenty years, four lines of research have expanded the conceptual and methodological approaches to studying the complexities of domestic work. The first tradition is that exemplified by Blood and Wolfe's study of marriage (1960), in which wives were asked who does more—husband or wife—on a list of half a dozen home chores. Presser's recent study (1977) continues this approach by asking wives if their husbands ever did any of a list of home chores. Moreover, in the last decade, a number of small surveys and case studies with housewives has created some of the first detailed social science accounts of the substance and process of contemporary home chores and their interrelatedness to child rearing (e.g., Berheide and Berk 1977, as mentioned above; Komarovsky 1964; Lopata 1971; Oakley 1976; and Piotrkowski 1979).

Third, several important time-budget studies have collected detailed time-diaries on all daily activities from large populations of men and women, thus providing the first large-scale quantitative information on what domestic activities people did, who did them, how much time they took, and what activities were done simultaneously (Walker and Woods 1976; Szalai 1972; Meissner et al. 1975; Robinson 1977a, 1977b). Finally, the Survey Research Center of the University of Michigan Institute for Social Research has developed several summary measures which ask respondents to estimate the number of hours they and their spouses spend on home chores and child rearing. These questions are conceptually similar to the summary time measures used extensively in research on work in which subjects are asked how much time they spend on breaks, on overtime, and so forth. The information from such questions is not simply the kind found in employment contracts, but reflects individuals' more subjective assessments of how they actually spend their time (Staines et al. 1978; Pleck et al. 1978).

Subsequent research revealed that Blood and Wolfe's percentage approach to measuring family work had two major limitations: first, it could not capture shifts in relative family work contributions occurring as a result of an absolute reduction in one partner's activity, thereby reducing the total spouse time on family work; and second, without reports of minutes and hours, comparisons could not be made between families about actual amounts of time in particular activities.

These methodological problems came to light through the time-budget research when it became apparent that husbands did not appreciably increase their participation in family work when their wives were employed—as the other methodology had suggested; rather, the wives simply decreased their total time, thus increasing the husband's time proportionately, even when their actual time was not increased at all, or only marginally.

Although the "summary" items cannot provide the detailed and longitudinal information of the time-budget diaries, they are useful when the field setting for the research allows only measures which take a short time for respondents to answer—as in the self-administered questionnaires used in the government agencies in this survey. This field-constraint was demonstrated in an early pre-test for this study. Twenty subjects were asked to complete two different kinds of family work questions. One set of questions, the Michigan items, asked respondents to estimate the average total time they spent daily on all home chores and child rearing. The second set separated out the list and asked respondents how much time they spent daily on each chore. The first form was answered quickly and easily by both sexes. The second form of the question produced irritation and some refusals to answer the questions in complete and useable form. Males were more negative than females.

The explanation for the different effectiveness of the two measures in self-administered questionnaires seems related to aggregation and regularity issues. Apparently the best level of aggregation for assessing time spent on family work is one which asks people about categories of activities that they think of as being done on a regular or uniform basis. For example, if people do laundry once or twice a week, it is difficult to calculate how much time they spend doing laundry on a daily basis; but if they know they spend a couple of hours a day doing any of a variety of household chores, it is relatively easy to estimate the average amount of time on these as a group. (This interpretation was suggested by Graham L. Staines.)

For these reasons, the Michigan items were chosen for this study. The home chores items ask, "On the average, on days when you are working, how much time do you spend on home chores in your family—things like cooking, cleaning, repairs, shopping, yardwork, keeping track of money and bills, and other work that has to be done around the house?" (Survey Research Center 1977a, p. 14). (Respondents answered in numbers of hours.) The same question is asked about non-working days and for spouses. The list of particular chores was developed from the most reliable items in a number of previous studies, and is intended to include some chores traditionally done by men, as well as those more typically done by women.

The child rearing question asks a parent to estimate how much time (in hours) he or she spends with the children each day doing any of the following: feeding, dressing, washing, going places, helping with homework or projects, disciplining, talking, reading, or chauffering. The question is asked for working and non-working days. The respondent is asked for similar estimates of his or her spouse's time.

For the present study weekly time spent on home chores for respondents was calculated by adding the number of hours reported on working days and non-working days, introducing a weighting factor of 5 for working days and of 2 for non-working days. The same procedure was used for time spent on child rearing. However, other problems with this methodology still remain. For example, if people spend more time on chores on working days, should this fact have any effect on the time they spend on non-working days? The question seems unanswerable, in theoretical terms, for some of the reasons discussed in Myrdal and Klein (1956), Berheide and Berk (1977), Oakley (1976) and others; in short, housework and other domestic enterprises have no universal norms or finite limits in most families; home improvements, as well as routine cleaning activities, for example—or creative meal-making or decor projects—may engage and expand people's time indefinitely. Personality differences may point to substantive differences in choices of activities without suggesting any ways of predicting actual time spent.

However, as the time-budget data from the 1972 twelve-nation study suggests, in all the countries studied, the sheer magnitude of family work that employed women feel they have to do makes their situation very different from employed men and housewives:

> Most striking of all are the work patterns that appear across all sites for employed women on weekends. Employed men turn some attention to various house maintenance and gardening chores over the weekend, but still indulge themselves in relatively large amounts of leisure activity. The housewife also shows a universal tendency to employ Sunday as a day of rest: while cooking is a residual necessity, her time spent on housework generally falls off almost 50 percent on that day. The employed woman, however, just about doubles the amount of time spent on housework on her days off from work: clearly she must use them to catch up on these obligations, rather than profit from them for rest and recuperation.
> Robinson, Converse, and Szalai 1972, p. 121.

Finally, individuals' daily estimates of their time use (as opposed to smaller unit time-diary information) are likely to be crude for at least two additional reasons: first, most of the family work asked about takes place in smaller than hourly segments of time; indeed it is characterized by a few minutes on one activity and a few on another. Therefore, when a respondent guesses about an hourly total, he or she necessarily rounds off the minutes very roughly. Second, the rounding-off is likely to be upward or downward depending on social desirability issues. For "time with children" parents may exaggerate the time, mothers especially, because they think they ought to spend more time with their children. Conversely, some males may feel reluctant to admit they spend much time on home chores, while others may want to appear to spend more time. At this period of changing social values on this topic, it is difficult to know which direction the "social desirability" factor cuts for males.

A similar procedure was used to determine the time spent on home chores and on child rearing for spouses, the only difference was in the weighting factor which was dependent on the number of days worked by the spouse.

Family Equity

To measure equity in family work between husbands and wives, the total hours reported for both spouses on child rearing and home chores were summed and the percentages were calculated for each partner. Two separate additional questions also asked who had the main responsibility for day-to-day arrangements for and care of the children, and who had the main responsibility for seeing that home chores got done.

Appendix E

Effects of asymmetry in male and female family roles

The following discussion supplements the "equity" section of the text (see Chapter 4) by outlining three additional ways in which the "asymmetry" of male and female roles in family and work may have negative effects on the well-being of various family members. Although the three issues are not used as direct outcome measures for the survey, all of them are reasons why various advocates of flexitime hope that males and females will increasingly have more "symmetrical" roles in work and family.

Employment Opportunity Costs for Women

In addition to the statements supporting equal employment opportunity for women, quoted from the congressional hearings on alternate work schedules (Chapter 3), a number of studies have demonstrated that when conflicts between family and employment needs arise in families, women characteristically make choices in favor of the family. Men, on the other hand, put the job first, as long as there is a woman to manage the children's needs—as there usually is. Only .01 percent of American fathers bring up their children without mothers (Current Population Reports 1977); and only 1.5 percent of American children are raised by fathers alone (Bureau of Labor Statistics 1978, Table 4).

For example, when someone must stay home with sick children who cannot go to school or to a babysitter, or when someone must take children to a medical appointment, or should attend a school or sport performance, or meet with a teacher, or drive a child somewhere after school, or shop or prepare dinner, or put the children to bed, almost invariably the mother, employed or not, arranges her schedule to do these things—as well as fitting in routine errands, home chores, and the like. These familiar interruptions in the job efforts of employed women—to say nothing of the months or years taken out of the labor market for child bearing and rearing—are reflected in the employment status and wage gaps between males and females, and in employers' attitudes towards hiring and promoting women (Rapoport and Rapoport 1971; Holmstrom 1972; Epstein 1970; Poloma and Garland 1971; Berheide and Berk 1976; Ralph E. Smith 1977b, 1977c, 1979).

With respect to the focus of this study, structural changes in employment, such as wider scheduling options through part-time and flexitime arrangements, are seen as partial solutions to managing the conflicts of work and home responsibilities. But the critical issue, in terms of sex role equity in employment and family roles, is whether such employment options are used only or largely by women, thus perpetuating the interlocking sequence in which women seek, are offered, and take less demanding jobs than males—and therefore do not move as quickly up the occupational hierarchies—because of their family responsibilities. (Another side of the issue, not investigated here, is whether flexitime options encourage more mothers to be employed.)

Once people are given a choice in work scheduling, the next important question is, of course, how will they use the time? At this level the issue becomes one of what they believe is important for them to do, or what they are expected to believe (e.g., Clark, Nye, and Gecas 1978; Polatnik 1973). Coser discusses the matter in terms of women's professional employment in the following way:

> The conflict is one of allegiance, and it does not stem from the mere fact of involvement in more than one social system. Such conflicts do not typically arise in the case of husbands. Men can be fully engaged in their occupations without fear of being accused of a lack of devotion to the families. It is only when there is a normative expectation that the family will be allocated resources of time, energy, and affect that cannot be shared with other social institutions that conflict arises. And this typically occurs only in the case of women who have the cultural mandate to give primary allegiance to their families. Hence, this mandate sharply limits the access of women to high-status positions and skews the distribution of power in

> the family in the direction of the male head of the household. This is why the notion of equal access of women to high-status positions in American society is presently discussed with such great affect [*sic*]. What is at stake is not so much equal access to job opportunities as such, but equal power within the family. Power depends on resources, and women who do not have occupational resources are in a very poor position to share it equally with their husbands.
>
> Coser 1974, p. 92

This opportunity-cost issue is not measured directly in this study, but two expectations are related to it in terms of flexible work scheduling; first, that with greater schedule flexibility women will be able to attend to family needs without jeopardizing their work status; and second, that men with greater schedule flexibility will share the family activities with them more equally. (Assessment of the opportunity-cost issue could only be determined in a longitudinal study with the same subjects.)

Effects on Children

The effects on children of inequities in their mothers' and fathers' roles in family and work have been assessed from various standpoints. The multitude of twentieth-century research on various aspects of child development reflects the pendulum swings in attitudes and practices—for example, between permissive and authoritarian notions of parental treatment of children. Heretofore, the work-time/family-time issues have been addressed in research and policymaking primarily in terms of prevailing assumptions about the importance of mother-child relationships (e.g., Maccoby 1958). As Hoffman and Nye put it: "The most persistent concern about maternal employment has to do with the absence of the mother from the home while she is working and the consequent fear that the child then lacks supervision, love, and cognitive enrichment" (Hoffman and Nye 1974, pp. 147–48).

Also, as several cross-cultural studies have documented, nonemployed American mothers typically spend much more time with their children than mothers in other societies (e.g., Minturn and Lambert 1964; Whiting and Whiting 1975). Since Freidan's (1963) attack on "the feminine mystique," a number of empirical and theoretical studies have discussed the negative dimensions of full-time housewifery and mothering, suggesting that children's interests may not be best served by full-time mothers who feel isolated and overworked, yet under-challenged, in the exclusive and omnipresent company of their young. Typically, full-time mothers get little help from their husbands in the "continuous process" jobs of household management and child rearing, as demonstrated in the time-budget and other research (e.g., Bernard 1974; Bronfenbrenner 1977a; Fidell 1976; Henry 1965; Hoffman and Nye 1974; Oakley 1976; Piotrkowski 1979; Rapoport et al. 1977; Rich 1976; Safilios-Rothschild 1974; Seidenberg 1975; Wortis 1974; and Wright 1978).

A substantial body of research on fathers documents the fact that they play peripheral roles in the family, compared to mothers, in terms of the everyday parenting aspects of family life. Recent research on "father absence" refers not only to fathers whose travel keeps them away from home, but to those who are home every night but with minimal sharing in family work or interaction with family members. Children whose fathers opt for increased income and job advancement versus family involvement are increasingly seen as "paternally deprived" (Rapoport et al. 1977; Siegman 1966; Shinn 1977; Lynn 1974; Biller 1976; Gerzon 1970; Lamb 1976). With respect to cognitive development per se, in her review of research on intact two-parent families, Clarke-Stewart (1977, p. 53) reports that "children whose fathers are home more than two hours a day are academically superior to those whose fathers are home less than one hour a day." In another review of fathers' roles, Hetherington concludes: "The father with low participation in the family or with low warmth in his relationship with his family may be just as detrimental to the child's intellectual growth as one who is totally absent. . . . The presence of the father is not the important variable; the important variable is the participation of a good father" (Hetherington et al. 1977, p. 9).

In a study of the role of parents in the sexual learning of children, Roberts, Kline, and Gagnon (1978) reported:

> Most fathers do not spend much time with their children, particularly their young children, and this certainly makes it difficult for them to take as active a role as mothers in verbal communication. . . . In those few households where fathers share and participate equally in general household and child care tasks, both parents reported that the child is more will-

> ing to ask the father questions, and the likelihood of father-child dialogue is greatly increased [p. 79].
>
> It seems clear that if the child sees the father as more involved in the family, and if the wife sees the father as more active in sharing household tasks, the father, in fact, becomes more "askable" to his children and his presence in family discussion is seen as more desirable by his wife [p. 74].

A significant shift in expectations for fathers' roles is reflected in recent scholarly and how-to-parent writings. In his most recent book for parents, Spock (1976) asserts that fathers should be involved in the "nitty-gritty" of parenting chores, whether or not the mother is employed. In his new conception, home chores—like shopping, cooking, cleaning and laundry—are considered part of parenting for fathers, as well as for mothers. Summing up a new orientation towards child rearing reflected in current child care journal articles, De Frain (1974) reports agreement on the following propositions: (1) that the burdens of child care are too great for mothers alone; (2) benefits of child rearing are too great to be the sole possession of mothers; and (3) children are the joint responsibility of both parents (Rapoport et al. 1977, p. 57; Levine 1976).

Most empirical research on this issue suggests that when men do increase their participation in domestic work, the increments appear in activities traditionally considered masculine, or they concentrate on the presumably more enjoyable activities, such as playing with children (e.g., Blood and Wolfe 1960; Maklan 1977; Oakley 1976; Johnson and Johnson 1977; Racki 1975). Two recent studies with information on fathers on shift work suggest that when, by necessity—i.e., the mother is at work—fathers take over the full range of child rearing responsibilities for a substantial portion of the day, they become closer to their children (Hood and Golden 1979; Lein et al. 1974).

In this research we do not directly assess the effects of equity in mother-father roles on children, but we measure the extent of equity with these above concerns in mind. We accept, hypothetically, the implication of these research findings, that the closer to equity parents can come in terms of their time with their children, the better for the children. The question for our study is whether fathers on flexitime report sharing more time with their children than their counterparts on standard time.

Effects on Marital Satisfaction

To date, little research has investigated whether greater symmetry in family work increases marital satisfaction, largely because there has been little symmetry to examine and little interest in the issue until recently. However, following the expectations of the preceding discussion—that greater equity in family work would bring other advantages to husbands, wives, and children—the implication is that marital relationships would also benefit from greater equity in family work.

Analyses of one set of British data suggest the kinds of advantages. In her study of careers and marital happiness, for example, Bailyn (1970) concludes that in both conventional and dual career families, happiness is greater when the differences between husband and wife in family-related interests and activities are minimized. In particular, happy dual-career marriages are more likely "when the wife is not responsible alone for . . . the house and the children" (pp. 106–107). Analyzing data from the same two hundred fifty British couples, the Rapoports and Thiessen (1974) conclude that both husbands and wives found more enjoyment in a variety of everyday activities when the husband reported that he was "family-oriented." Although this study of flexitime makes no attempt to gauge the marital satisfaction of respondents directly, the interest in equity in family work derives, in part, from this line of thinking.

Appendix F

Correlations among dependent variables

As discussed in Chapter 4, our study has three major outcome variables: stress, time (spent in family work), and the sharing of family work between spouses. Each of these has more than one indicator. The stress variable has two scales as indices of stress; the time variable is measured in terms of time with children, time on chores, and how the total time on these activities is divided between spouses (in husband-wife families).

The amount of correlation among the indicators corroborates the intention that the stress items and time items should measure different variables; that is, the stress indicators (the two scales) are significantly correlated with each other at .46 ($p \leq 0.00$); and the time indicators (time with children and time on home chores) are significantly correlated at .38 ($p \leq 0.000$). Similarly, the time and stress indicators are not highly correlated with each other, thus indicating that they are tapping different constructs (see Table 11).

For example, although the correlations between the Role Strain Scale and the two time indicators are significant and in a positive direction, they are quite low: .11 for the time with child rearing and .17 for the time on home chores. For the Family Management Scale, on the other hand, the correlations are negative: −.20 for time with child rearing ($p \leq .002$); and −.03 for the time on home chores, although this correlation is not significant ($p \leq .3$).

These small and negative correlations between the stress and time indicators are consistent with expectations. In terms of the Role Strain Scale, it was anticipated that more time spent on home chores would

TABLE 11
PEARSON CORRELATION COEFFICIENTS AMONG DEPENDENT VARIABLES

Characteristics	*Adult Role Strain*	*Total Role Strain*	*Adult Manage-ment*	*Total Manage-ment*	*Time on Child Rearing*	*Time on Home Chores*
Adult Role Strain Scale	1.00		0.46 $p \leq 0.0001$			0.17 $p \leq 0.0001$
Total Role Strain Scale		1.00		0.46 $p \leq 0.0000$	0.11 $p \leq 0.0360$	0.17 $p \leq 0.0030$
Adult Management Scale			1.00			0.10 $p \leq 0.0300$
Total Management Scale				1.00	− 0.20 $p \leq 0.0020$	− 0.03 $p \leq 0.3000$
Time on Child Rearing					1.00	0.38 $p \leq 0.0000$
Time on Home Chores						1.00

result in more strain. The smaller correlation between child rearing and the Role Strain Scale is a more ambiguous finding. The study hypothesized that the more time a person spent on child rearing the lower the role strain would be, but since none of the other factors that would affect the amount of time in child rearing were controlled for (such as number of children, ages of children, outside help, sex, and earning status of spouse) in determining these correlation coefficients, the meaning of the correlation figure itself is unclear for these two variables. On the other hand, the negative correlations between the Management Scale and the time measures are consistent with our expectations in the sense that the Management Scale focuses on logistical factors in family life that are not primarily concerned with the actual amounts of time spent on children and chores (see Table 11 for these correlations).

Appendix G

Government service job classifications (PATCO)

The federal service white collar occupational classifications referred to in Chapter 5 are as follows.

1. *Professional occupations.* These require knowledge in a field of science or learning characteristically acquired through education or training equivalent to a bachelor's or higher degree with major study in, or pertinent to, the specialized field, as distinguished from general education. The work of a professional occupation requires the exercise of discretion, judgment, and personal responsibility for the application of an organized body of knowledge that is constantly studied to make new discoveries and interpretations, and to improve data, materials, and methods.

2. *Administrative occupations.* These involve the exercise of analytical ability, judgment, discretion, personal responsibility, and the application of a substantial body of knowledge of principles, concepts, and practices applicable to one or more fields of administration or management. While these positions do not require specialized educational majors, they do involve the type of skills (analytical, research, writing, judgment) typically gained through a college level general education, or through progressively responsible experience.

3. *Technical occupations.* These involve work typically associated with, and supportive of, a professional or administrative field that is non-routine in nature; such occupations involve extensive practical knowledge, gained through on-the-job experience and/or specific training less than that represented by college graduation. Work in these occupations may involve substantial elements of the work of the profes-

sional or administrative field, but requires less than full competence in the field involved.

4. *Clerical occupations.* These involve structured work in support of office, business, or fiscal operations; duties are performed in accordance with established policies, experience, or working knowledge related to the tasks to be performed.

5. *Other occupations.* These cannot be related to the above professional, administrative, technical, or clerical groups (Civil Service Commission 1976).

Appendix H

Sample questionnaire

The questionnaires for both agencies were identical in all but the following respects:

- They were printed on different colored paper.
- The direct questions on flexitime, numbered 45 to 58 in the following copy of the instrument, were not included in the standard time agency version of the questionnaire.
- The standard time questionnaire included the following final item which did not appear on the flexitime version:

(45a) How important do you feel your job is compared to your spouse's job?
1. More important
2. Equally important
3. Less important
4. Not applicable

(45b) Why? ______________________________________
____
____

INSTITUTE FOR
EDUCATIONAL
LEADERSHIP

July 30, 1979

THE
GEORGE
WASHINGTON
UNIVERSITY

Suite 310
1001 Connecticut Avenue, N.W.
Washington, D.C. 20036

Samuel Halperin
Director
(202) 676-5900

Education Policy Fellowship Program
(202) 676-5925

Educational Staff Seminar
(202) 676-5949

The Associates Program
(202) 676-5935

"Options in Education" over National Public Radio
(202) 785-6462

Education of the Handicapped Policy Project
(202) 676-5910

Family Impact Seminar
(202) 296-5330

Washington Policy Seminar
(202) 676-5940

Fellowships in Educational Journalism
(202) 676-5901

Expanding Opportunities in Educational Research
(202) 676-3970

National Policy Fellows in Education of the Handicapped
(202) 676-5910

Dear

Last October you very kindly filled out the questionnaire for our study on work and family relationships. As you may remember, we mentioned in the cover letter that we would contact a few people again at a later date to talk with us further about these issues in their own lives. I am writing now to invite you and your spouse to participate in such a meeting.

You are invited for dinner and a discussion on Monday, August 6 from 5:30 to 9:30 p.m. at 4602 North Park Avenue, Chevy Chase, Maryland (see enclosed map). Three other Commerce Department employees and their spouses will be in the group. All of you are parents of young children. We are interested in learning more from you about how employed parents with young children handle their work and family responsibilities: What are the benefits? What are the difficulties? With your permission, we will videotape the discussion for our use in reviewing the issues which emerge from your comments. We will pay you $25 for any babysitting and transportation expenses you may incur. To insure your privacy and the confidentiality of your participation, only the research staff will see the tapes and your name will not appear in any reports of the study.

I will call you within the next few days to see whether you and your spouse can join us on August 6th. Both the discussion and the dinner (sandwiches) will be informal and we expect that you will find the occasion interesting and enjoyable. Thanks very much in advance for your help. I look forward to talking with you further.

Yours sincerely,

Halcy Bohen
Director, Work Schedules Study
Family Impact Seminar

HB/me

Enclosed: Map showing location of discussion
New York Times article describing the Family Impact Seminar

A Program of The George Washington University's Institute for Educational Leadership

To employees of the Maritime Administration

The attached questionnaire is for the Family Impact Seminar study on the effects of work on families, as described in the September 12th memorandum which you received from Russell Stryker. With the encouragement of the Maritime Administration, we are inviting you to participate in the study by answering the questions-- which should take about 30 minutes. To the best of our knowledge, this is the first survey which seeks to learn directly from U.S. Government employees about the effect of work on their family lives.

We anticipate that the information from the questionnaires will be very useful as the Government considers the effects its policies have on the families of its employees. We hope very much you will share your experiences so that the results of the survey will be comprehensive, accurate and timely.

Each questionnaire is numbered to facilitate the statistical analyses and because we would like to contact a few people again for interviews with their families. The information you give will be completely confidential and will never be associated with your name. To insure the confidentiality of your answers, please do not write your name on the questionnaire. You may decline to answer questions which do not seem reasonable to you, but the validity of the findings will be greatly enhanced if all questions are answered completely. Feel free to add any comments you may have in the margins.

Thank you very much for your interest and time in helping to make this study possible. If you have any questions while filling out the questionnaire, you are welcome to call Halcy Bohen at George Washington University (466-6133). A copy of the final report will be available for you to read in the Maritime Administration Office of Personnel in the Spring of 1979. Hopefully, the findings will have long-range benefits for all kinds of American families.

A. Sidney Johnson III
Director, The Family Impact Seminar
The George Washington University

Halcy Bohen
Director, The Work-Family Study
The George Washington University

Family Impact Seminar

SINCE WE USE THE WORD "FAMILY" THROUGHOUT THE QUESTIONNAIRE, WE THOUGHT YOU WOULD FIND IT HELPFUL TO KNOW OUR DEFINITION. WE START WITH THE CENSUS BUREAU DEFINITION -- "A FAMILY IS A GROUP OF TWO OR MORE PERSONS RESIDING TOGETHER WHO ARE RELATED BY BLOOD, MARRIAGE, OR ADOPTION" -- BUT WE ALSO INCLUDE RELATIVES WHO MAY NOT LIVE TOGETHER. FOR EXAMPLE, IF YOUR PARENTS, SIBLINGS OR CHILDREN ARE LIVING ELSEWHERE, BY OUR DEFINITION, THEY ARE PART OF YOUR FAMILY. OR IF YOU DO NOT HAVE CHILDREN, BUT HAVE SIBLINGS OR PARENTS (STILL LIVING), BY OUR DEFINITION, YOU ARE STILL PART OF A FAMILY.

TO ANSWER THE FOLLOWING QUESTIONS, PLEASE CIRCLE THE NUMBERS OF THE ANSWERS WHICH APPLY TO YOU, AND/OR FILL IN THE INFORMATION REQUESTED.

SECTION A

(1) Sex:

1. Male
2. Female

(2) Age:

1. 21 years and under
2. 22-25 years
3. 26-30 years
4. 31-35 years
5. 36-40 years
6. 41-50 years
7. 51-60 years
8. 61 years and up

(3) Present marital status:

1. Never married
2. First marriage
3. Remarried
4. Divorced
5. Widowed
6. Other (specify) ____________

(4a) How may children do you have? ____________

(4b) Please list their ages. ____________

(5) Educational background:

You:
1. High school or less
2. Some college
3. College degree
4. Advanced degree(s) Which?____________

Your spouse:
1. High school or less
2. Some college
3. College degree
4. Advanced degree(s) Which?____________
8. Not applicable

(6) What do you consider to be your ethnic and racial heritage?

1. American Indian or Alaskan Native
2. Asian or Pacific Islander
3. Black, not of Hispanic Origin
4. Hispanic
5. White, not of Hispanic Origin

(7) What kind of work do you do?
(For example, administrator of grants, typist, transportation specialist, management analyst, keypunch operator, etc.)

____

(8) What kind of position do you have? PLEASE CIRCLE THE ONE ALTERNATIVE WHICH BEST DESCRIBES MOST OF THE WORK YOU DO.

1. Professional
2. Administrative
3. Technical
4. Clerical
5. Other (Specify) ____________

(9) What is your service grade? ____________

(10) How satisfied are you with:

	Very satisfied	Satisfied	Neither Satisfied nor dis-satisfied	Dis-satisfied	Very dissatis-fied
a. your job in general	1	2	3	4	5
b. your pay	1	2	3	4	5
c. the number of hours you work	1	2	3	4	5
d. the schedule of your working hours	1	2	3	4	5
e. the sorts of things you do on your job	1	2	3	4	5

(11) How much do you agree or disagree with the following statements? CIRCLE THE NUMBER THAT BEST EXPRESSES YOUR VIEW FOR EACH ITEM.

	Strongly Agree	Agree	Neither Agree nor Disagree	Disagree	Strongly Disagree
a. What I do at work is more important to me than the money I earn.	1	2	3	4	5
b. My main satisfaction in life comes from my work.	1	2	3	4	5
c. It is much better for everyone involved if the man earns the money and the woman takes care of the house and children.	1	2	3	4	5
d. My main interest in my work is to get enough money to do the other things I want to do.	1	2	3	4	5
e. I'd be happier if I did not have to work at all.	1	2	3	4	5
f. A mother who works outside the home can have just as good a relationship with her children as a mother who does not work.	1	2	3	4	5

(12) About how many hours do you work on this job in the average week?

__________ hours

(13) What are your usual starting and ending hours?

__________ to __________

(14a) About how long does it usually take you to get from home to work, door to door?

_____________ minutes (one-way)

(14b) About how long does it usually take you to get from work to home, door to door?

_____________ minutes (one-way)

(15) Besides your main job and your family, do you have any other regular activities away from home (e.g., part-time jobs, evening courses, volunteer activities, etc.)? DO NOT COUNT SPECIAL SUMMER ACTIVITIES.

1. Yes. What? ______________________________

How many hours a week (on the average)? ____________

2. No.

(16) Who lives with you? Please list by relationship and age.

Example:

Relationship	Age
son	4
mother-in-law	55
spouse	36
niece	17
friend	25

Relationship	Age
____________	____
____________	____
____________	____
____________	____
____________	____
____________	____
____________	____
____________	____
____________	____

(17) Is there anyone (whether living with you or not) for whom you provide special care due to things like illness, a handicap, or old age?

1. Yes. Who? ____________

____

2. No.

(18) About how much of the money to support your family do you provide?

1. All
2. More than half
3. About half
4. Less than half

(19) On days when you are working, how easy or difficult is it for you to arrange your time to do each of the following? CIRCLE THE RELEVANT NUMBER FOR EACH ACTIVITY.

	Very Easy	Somewhat Easy	Neither Easy nor Difficult	Somewhat Difficult	Very Difficult	Not Applicable
To avoid the rush hour	1	2	3	4	5	8
To go to work a little later than usual if you need to	1	2	3	4	5	8
To go to health care appointments	1	2	3	4	5	8
To go on errands (e.g., shoe repair, post office, car serviced)	1	2	3	4	5	8
To go shopping (e.g., groceries, clothes, drug store)	1	2	3	4	5	8
To make telephone calls for appointments or services	1	2	3	4	5	8
To take care of your household chores	1	2	3	4	5	8
To help or visit neighbors or other friends	1	2	3	4	5	8
To participate in community activities	1	2	3	4	5	8
To adjust your work hours to the needs of other family members	1	2	3	4	5	8
To have meals with your family	1	2	3	4	5	8
To spend fun or educational time with your family	1	2	3	4	5	8

(20) Please indicate by circling the relevant number next to each statement how often you feel each of the following.

	Always	Most of the time	Some of the time	Rarely	Never	Not applicable
My job keeps me away from my family too much.	1	2	3	4	5	8
I feel I have more to do than I can handle comfortably.	1	2	3	4	5	8
I have a good balance between my job and my family time.	1	2	3	4	5	8
I wish I had more time to do things for the family.	1	2	3	4	5	8
I feel physically drained when I get home from work.	1	2	3	4	5	8
I feel emotionally drained when I get home from work.	1	2	3	4	5	8
I feel I have to rush to get everything done each day.	1	2	3	4	5	8
My time off from work does not match other family members' schedules well.	1	2	3	4	5	8
I feel I don't have enough time for myself.	1	2	3	4	5	8
I worry that other people at work think my family interferes with my job.	1	2	3	4	5	8
I feel more respected than I would if I didn't have a job.	1	2	3	4	5	8

(21) In general how satisfied are you with yourself as a:

	Very Satisfied	Satisfied	Neither Satisfied nor dissatisfied	Dissatisfied	Very Dissatisfied	Not applicable
1. parent	1	2	3	4	5	8
2. spouse	1	2	3	4	5	8
3. worker	1	2	3	4	5	
4. person	1	2	3	4	5	

IF YOU ARE NOT LIVING WITH A SPOUSE, BUT HAVE CHILDREN UNDER 18 YEARS OF AGE LIVING WITH YOU, PLEASE SKIP TO SECTION B.

IF YOU ARE NOT LIVING WITH A SPOUSE AND DO NOT HAVE CHILDREN UNDER 18 YEARS OF AGE LIVING WITH YOU, PLEASE SKIP TO QUESTION 43.

(22) Is your spouse employed at present? 1. Yes 2. No (SKIP TO QUESTION 29)

(23a) What kind of work does your spouse do?
(For example, aircraft mechanic, lawyer private practice, secretary, retail clerk, messenger Census Bureau, teacher, dental technician, etc.)

____

(23b) What kind of position does your spouse have? (CIRCLE ONLY ONE ALTERNATIVE)

1. Professional
2. Administrative
3. Technical
4. Clerical
5. Other (specify) ____

(24) What are your spouse's usual starting and ending hours at work?

____________ to ____________

(25a) About how long does it take your spouse to get from home to work, door to door?

________ minutes (one-way)

(25b) About how long does it take your spouse to get from work to home, door to door?

________ minutes (one-way)

(26) About how many hours does your spouse work on this job in the average week?

__________ hours

(27) Does your spouse have flexible working hours?

1. Yes (please describe)

____

2. No

(28) How much do you think your spouse would agree or disagree with the following statements for himself or herself?

	Strongly Agree	Agree	Neither Agree nor Disagree	Disagree	Strongly Disagree
a. What I do at work is more important to me than the money I earn.	1	2	3	4	5
b. My main satisfaction in life comes from my work.	1	2	3	4	5
c. My main interest in my work is to get enough money to do the other things I want to do.	1	2	3	4	5
d. I'd be happier if I didn't have to work at all.	1	2	3	4	5

(29) Does your spouse have any other regular activities away from home besides his/her main job and family responsibilities (e.g., part-time jobs, evening courses, volunteer activities, etc.)? DO NOT COUNT SPECIAL SUMMER ACTIVITIES.

1. Yes. What? ______________________________________

____

How many hours a week on the average? ____________

2. No.

SECTION B

THE FOLLOWING GROUP OF QUESTIONS IS RELATED TO THE HOME CHORES IN YOUR FAMILY--THINGS LIKE COOKING, CLEANING, REPAIRS, SHOPPING, YARDWORK, KEEPING TRACK OF MONEY AND BILLS--PLUS PLANNING AND ARRANGING FOR ALL THAT HAS TO GET DONE.

(30) In your household, who has the main responsibility for seeing that the kinds of home chores listed above get done?

1. I have the main responsibility.
2. My spouse and I share the responsibility equally.
3. My spouse has the main responsibility and I help out.
4. Other arrangements. What? ______________________

(31) Do you hire anyone from outside to help with the home chores on a regular basis?

1. No.
2. Yes. What do they do? ______________________

How many hours per week? ______________________

(32) How many hours a day do each of the following persons spend on home chores like those listed above?

	On days when working	On days when not working	Not Applicable
you			
your spouse			
your children	(on school days)	(on non-school days)	
others in the household Who? ________			

(33) How much time would you like your spouse to spend on home chores?

1. More time than now
2. Less time than now
3. Same amount as now
8. Not applicable

(34) How much time would your spouse like you to spend on home chores?

1. More time than now
2. Less time than now
3. Same amount as now
8. Not applicable

IF YOU DO NOT HAVE CHILDREN UNDER 18 YEARS OF AGE LIVING WITH YOU, PLEASE SKIP TO QUESTION 43.

SECTION C

THE FOLLOWING GROUP OF QUESTIONS ARE RELATED TO THE TIME YOU SPEND TAKING CARE OF OR DOING THINGS WITH YOUR CHILDREN--THINGS LIKE FEEDING, DRESSING, WASHING, GOING PLACES, HELPING WITH HOMEWORK OR PROJECTS, DISCIPLINING, TALKING, READING, DRIVING THEM PLACES, ETC.

(35) On the average, how much time do you spend on any or all of the above?

On days when you are working? ____________ hours per day

On days when you are not working? ____________ hours per day

(36) In your family, who has the main responsibility for the day-to-day arrangements for and care of the children?

1. I have the main responsibility.
2. My spouse and I share the responsibility equally.
3. My spouse has the main responsibility and I help out.
4. Other arrangements. What? ________________________

(37) In your family, in particular, what kinds of things do you do with your children?

Times per week

____

____

____

____

(38a) On the average about how much time do your children spend watching television?

____________ hours per week

(38b) On the average about how much time do you spend watching television with your children?

____________ hours per week

IF YOU ARE NOT LIVING WITH A SPOUSE, PLEASE SKIP TO QUESTION 41.

(39) On the average, how much time does your spouse spend taking care of and doing things with your children like those listed above under Section C.

On days when he/she is working? ____________ hours per day

On days when he/she is not working? ____________ hours per day

(40a) How much time would you like your spouse to spend taking care of or doing things with your children?

1. More time than now
2. Less time than now
3. Same amount as now
8. Not applicable

(40b) How much time would your spouse like you to spend taking care of or doing things with your children?

1. More time than now
2. Less time than now
3. Same amount as now
8. Not applicable

(41) Please indicate by circling the relevant number next to each statement <u>how often you feel</u> each of the following.

	Always	Most of the time	Some of the time	Rarely	Never	Not Applicable
I worry whether I should work less and spend more time with my children.	1	2	3	4	5	8
I am a better parent because I am not with my children all day.	1	2	3	4	5	8
I find enough time for the children.	1	2	3	4	5	8
I worry about how my kids are while I'm working.	1	2	3	4	5	8
I have as much patience with my children as I would like.	1	2	3	4	5	8
I am comfortable with the arrangements for my children while I am working.	1	2	3	4	5	8
Making arrangements for my children while I work involves a lot of effort.	1	2	3	4	5	8
I worry that other people feel I should spend more time with my children.	1	2	3	4	5	8

(42) On days when you are working, how easy or difficult is it <u>for you</u> to do each of the following? CIRCLE THE RELEVANT NUMBERS FOR EACH ACTIVITY.

	Very Easy	Somewhat Easy	Not Easy or Difficult	Somewhat Difficult	Very Difficult	Not Applicable
To take your children to health care appointments	1	2	3	4	5	8
To take your children to or from a child care setting or school	1	2	3	4	5	8
To go places with your children after school	1	2	3	4	5	8
To go to school events and appointments for your children	1	2	3	4	5	8
To make alternative child care arrangements when necessary (e.g., school snow days)	1	2	3	4	5	8
To be home when your children get home from school	1	2	3	4	5	8
To stay home with a sick child	1	2	3	4	5	8
To make arrangements for children during summer vacations	1	2	3	4	5	8
To have relaxed, pleasant time with your children	1	2	3	4	5	8

(43) How much do your job and family life interfere with each other?

1. Not at all.
2. Not too much.
3. Somewhat.
4. A lot.

(44) In what ways do they interfere with each other?

____

____

____

SECTION D

THE FOLLOWING QUESTIONS ARE RELATED TO THE FLEXITIME ARRANGEMENTS IN THE MARITIME ADMINISTRATION

(45) What time and attendance methods does your office use?

1. We use an honor system.
2. We use a sign-in/sign-out sheet or board.
3. Other. What? ____________________

(46) Do you have to tell your supervisor in advance the hours you expect to work?

1. Yes
2. No

(47) Have you changed the schedule of your working hours since your agency offered flexitime?

1. No. I still work the same hours.
2. No. My agency had flexitime when I started.
3. Yes, in the following way:

	my hours used to be	my hours are now
starting time		
lunch time		
leaving time		

(48) Do any of the following limit you from using the flexitime program in your agency as much as you would want to? CIRCLE THE RELEVANT NUMBER FOR EACH ALTERNATIVE.

		Yes, a great deal	Somewhat	Not at all
1.	Carpool	1	2	3
2.	Parking	1	2	3
3.	Bus schedule	1	2	3
4.	Supervisors	1	2	3
5.	Need to meet with or coordinate my work with others	1	2	3
6.	Child care arrangements	1	2	3
7.	Spouse schedule	1	2	3
8.	Other limits. What? ____________	1	2	3

____

____

(49) How often do you vary your starting time by more than 15 minutes?

1. Frequently (more than twice a week)
2. Usually (about once or twice a week)
3. Occasionally (once every 2 weeks)
4. Never

IF YOU HAVE NOT BEEN ON STANDARD TIME BEFORE USING FLEXITIME, PLEASE SKIP TO QUESTION 52.

(50) Has flexitime enabled you to participate in any of the following types of activities to a greater extent than you could participate in them under fixed working hours?

a. Recreational activities

1. No.
2. Yes. What? ____________________

b. Educational Activities

1. No.
2. Yes. What? ____________________

c. Greater amount of time spent with family

1. No.
2. Yes. Doing what? ____________________

____

____

____

(51) Has your use of flexitime changed the amount of time it takes you to commute to and from work?

1. No.
2. Yes, it has reduced my commuting time.
   About how many minutes less round trip? ________ minutes
3. Yes, it has increased my commuting time.
   About how many minutes more round trip? ________ minutes

(52) How important for you are each of the following reasons for using flexitime? CIRCLE THE RELEVANT NUMBER FOR EACH REASON.

		Very Important				Not Important
1.	To commute when there is less traffic.	1	2	3	4	5
2.	To better schedule personal activities before and after work.	1	2	3	4	5
3.	To use less annual and sick leave for personal or family business.	1	2	3	4	5
4.	To have less worry about getting to work "on time."	1	2	3	4	5
5.	To use the quiet time in the office before or after core time to work.	1	2	3	4	5
6.	To have more time with your family.	1	2	3	4	5
7.	To get more rest.	1	2	3	4	5
8.	Other? ____________	1	2	3	4	5

(53) In general, how successful do you think flexitime in your agency has been?

1. Very successful
2. Successful
3. Only partially successful
4. Unsuccessful

(54) Why do you think so?

____

____

____

(55) Would you like any other kind of additional flexibility in your schedule? CIRCLE NUMBERS OF ANY YOU WOULD LIKE.

1. Longer flexible bands. What hours? ____________
2. A 4-day week with 10 hour days.
3. A 4-day week with proportionately less pay.
4. A banking, borrowing arrangement so I could work more or less than 8 hours a day, as long as I averaged 80 hours in each 2-week period.
5. More part-time opportunities. What type? ____________
   ____
6. Longer vacations.
7. Other variations? ____________
   ____

(56) Are you a supervisor?

1. Yes. How many people do you supervise? ____________
2. No. (SKIP TO QUESTION 58.)

(57) Do you think additional flexibility could be used in your office?

1. Yes. In what ways? ______________________________

____

2. No. Why? ______________________________________

(58)

Some people have found that more flexible work schedules have significantly affected their lives outside of work. If your personal and/or family life has been affected by greater flexibility in your work schedule, please describe these changes.

Many thanks for your time and interest in answering the questions.

Appendix I

Statistical techniques

The basic analysis of the data was done using difference of means tests (*t* tests) to compare the scores of all respondents in the two agencies on the stress scales, and on time spent on family work. Difference of means tests were also used to make the same comparisons for each of the family subgroups (see the figure at the end of Chapter 5.) A multivariate analysis was then used in order to take into account the possible influence of five other factors in addition to schedule: total hours worked weekly, total hours commuted weekly, family life cycle stage, occupational level, and outside help with family work.

The Multiple Classification Analysis (MCA) technique was used for the multivariate analysis (Andrews, Morgan, and Sonquist 1967). It is an option provided by the Statistical Package for the Social Sciences (SPSS; see Nie et al. 1970). For our study, the main advantages of the MCA method were the following. First of all, it can be used when the independent variables have not been experimentally manipulated, hence the independent and control variables may be correlated (e.g., in this study a moderate relationship exists between schedule, the independent variable, and the total number of hours worked, one of the control variables [$r = .17$]).

Second, it can handle categorical (i.e., nominal) independent variables (e.g., flexitime and standard time schedules). Finally, the computer results of the analysis are presented directly in an easily comprehensible format that shows how the means change when the controls are introduced. For example, the MCA table shows that the mean weekly hours in home chores for all single mothers is 26 hours and that standard time single mothers spend 4.5 hours less than this mean, while flexitime mothers spend 3.1 hours more than the mean. In contrast, in the dummy variable multiple regression technique, one category from each explanatory variable is omitted and coefficients for the remaining categories are expressed as deviations from the omitted category. For this study, the MCA version was more useful because (1) coefficients for all categories are obtained, and (2) the coefficients are expressed as deviations from the mean, which is a more easily understood form. This MCA technique was also used for analysis of the six-nation time-budget data (e.g., see Robinson 1977a, 1977b).

The MCA technique estimates a coefficient for each category of each explanatory variable in terms of deviations from the grand mean. These coefficients show how each explanatory variable is related to the dependent variable, both before and after adjusting for the effects of the other explanatory variables. In addition, Eta and Beta coefficients are computed as measures of the relative importance of each explanatory variable. Finally, a multiple correlation coefficient is computed which indicates the magnitude of the relationships between the dependent variable and all explanatory variables considered together. This multiple correlation coefficient squared indicates the amount of variance explained by the independent and control variables.

For example, when considering the effects that the flexitime schedule may have on job-family role strain—while controlling for family life cycle, sex, earning status, and occupational level—we can see the effects of each explanatory variable on the job-family role strain of respondents (expressed in deviations from the mean for the whole sample) before adjusting for the effects of the other variables. Second, we can see if flexitime still has an effect after controlling for all the other variables. In addition, we can see the strength of the relationships between job-family role strain and all the explanatory variables. Furthermore, through the squared multiple correlation coefficient (R^2) we can assess how much of the variance in the dependent variable can be accounted for by this specific set of explanatory variables.

Appendix J

Data tables

The numbers of respondents in the following subcategories on some of the following tables may not add to totals because some individuals did not provide information for some questions (e.g., family earning structure).

TABLE 12

COMPARISON OF MEAN SCORES FOR ADULT AND TOTAL FAMILY MANAGEMENT SCALE (SCALE RUNS FROM 1–5)*

	Standard Time			*Flexitime*			
Sample	*N*	*Mean*	*SD*	*N*	*Mean*	*SD*	*Significance Level*
Adult Family Management Scale							
Total sample	239	3.1	0.7	313	2.7	0.8	†
Parents	123	3.1	0.7	145	2.9	0.8	‡
Husband and wife families	98	3.1	0.7	111	2.9	0.8	‡
Women with employed husbands	42	3.1	0.8	24	3.2	0.6	
Men with employed wives	28	3.1	0.6	37	2.8	0.8	
Men with non-employed wives	28	3.3	0.7	49	2.8	0.8	‡
Single mothers	15	2.9	0.6	26	2.7	0.8	
Non-parents	116	3.1	0.8	163	2.6	0.8	†
Husband and wife families	56			80			
Women with employed husbands	20	3.5	0.6	30	2.6	0.7	†
Men with employed wives	25	2.9	0.8	35	2.8	1.0	
Men with non-employed wives	11	2.7	0.8	15	3.0	0.9	

Singles	52	3.2	0.7	85	2.5	0.7	†
Women	32	3.2	0.7	57	2.5	0.7	†
Men	20	3.3	0.8	28	2.4	0.7	†
Total Family Management Scale							
Parents	105	3.2	0.7	116	3.0	0.7	
Husband-wife families	83	3.2	0.7	84	3.0	0.7	
Women with employed husbands	38	3.1	0.8	22	3.2	0.6	
Men with employed wives	22	3.2	0.6	31	2.9	0.7	
Men with non-employed wives	23	3.5	0.7	31	3.0	0.8	‡
Single mothers	14	3.1	0.7	24	2.8	0.8	

*Higher scores = greater stress.
†Significant at $p \leq .01$.
‡Significant at $p \leq .05$.

TABLE 13
COMPARISON OF MEAN SCORES FOR ADULT AND TOTAL JOB-FAMILY ROLE STRAIN SCALE (SCALE RUNS FROM 1–5)*

	Standard Time			Flexitime			
Sample	*N*	*Mean*	*SD*	*N*	*Mean*	*SD*	*Significance Level*
Adult Job-Family Role Strain Scale							
Total sample	262	2.7	0.5	326	2.6	0.4	†
Parents	132	2.8	0.6	155	2.7	0.5	‡
Husband and wife families	106	2.9	0.5	121	2.7	0.5	†
Women with employed husbands	43	3.1	0.6	26	2.9	0.5	
Men with employed wives	31	2.7	0.4	40	2.6	0.5	
Men with non-employed wives	32	2.7	0.4	54	2.6	0.5	
Single mothers	15	2.8	0.6	26	2.7	0.7	
Non-parents	130	2.6	0.6	170	2.5	0.6	
Husband and wife families	66			84			
Women with employed husbands	20	2.9	0.5	25	2.5	0.9	‡
Men with employed wives	30	2.4	0.5	36	2.5	0.5	
Men with non-employed wives	16	2.4	0.5	23	2.5	0.5	

Singles	56	2.7	0.6	77	2.5	0.6	
Women	33	2.6	0.6	51	2.6	0.6	
Men	23	2.8	0.7	26	2.4	0.6	‡
Total Job-Family Role Strain Scale							
Parents	125	2.6	0.5	145	2.5	0.5	‡
Husband and wife families	101	2.7	0.5	112	2.5	0.4	‡
Women with employed husbands	41	2.8	0.6	24	2.8	0.4	
Men with employed wives	30	2.6	0.4	38	2.4	0.5	
Men with non-employed wives	30	2.5	0.3	50	2.4	0.4	
Single mothers	15	2.8	0.7	25	2.7	0.6	

*Higher scores = greater stress.
†Significant at $p \leqslant .01$.
‡Significant at $p \leqslant .05$.

TABLE 14
MEAN SCORES ON THE TWO STRESS SCALES FOR WOMEN AND MEN ON STANDARD TIME AND FLEXITIME*

	Adult Family Management Scale			*Adult Family Strain Scale*		
Sample	*N*	*Mean*	*SD*	*N*	*Mean*	*SD*
Women						
Standard time	118	3.1	0.7	120	2.8	0.6
Flexitime	142	2.7	0.8	140	2.7	0.7
Men						
Standard time	121	3.1	0.7	142	2.6	0.5
Flexitime	170	2.7	0.5	186	2.5	0.4

*The differences between the mean scores for both women and men are significant at $p \leq .05$ on both scales. Higher scores = greater stress.

TABLE 15

RESPONDENTS' TIME ON HOME CHORES (WEEKLY, FOR WORKDAYS AND FOR OFFDAYS).[†]

	Standard Time				Flexitime			
Sample	*N*	*Mean Weekly Hours*	*Mean Hours on Workdays*	*Mean Hours on Offdays*	*N*	*Mean Weekly Hours*	*Mean Hours on Workdays*	*Mean Hours on Offdays*
Total sample	192	17.8	1.6	4.9	229	19.3	1.7	5.4
Parents	123	20.2	1.8	5.5	146	20.5	1.8	5.6
Husband and wife families	100	19.9	1.8	5.3	118	18.7	1.6	5.4
Women with employed husbands	40	26.3	2.7	6.3	26	27.8	2.6	7.2
Men with employed wives	31	17.7	1.5	5.1	39	16.6	1.3	4.8
Men with non-employed wives	29	13.5	0.97	4.2	51	15.7	1.2	4.8
Single mothers	15	21.6*	2.0*	5.8	21	29.2*	3.1*	7.0
Non-parents	65	11.6*	0.9	3.6*	78	16.0*	1.3	4.9*
Women with employed husbands	20	18.5	2.0	4.4	21	20.3	2.1	4.9
Men with employed wives	28	11.5	0.9	3.5	35	15.3	1.2	4.7
Men with non-employed wives	17	11.9	0.9	3.6	22	15.2	1.1	4.9

*Significant at $p \leq .05$.

†This information was not collected for single non-parents.

TABLE 16
RESPONDENTS' TIME ON CHILD REARING*

	Standard Time				*Flexitime*			
Sample	*N*	*Mean Weekly Hours*	*Mean Hours on Workdays*	*Mean Hours on Offdays*	*N*	*Mean Weekly Hours*	*Mean Hours on Workdays*	*Mean Hours on Offdays*
Total parents	123	21.7	2.2	5.4	142	20.2	2.0	5.3
Husband and wife families	98	20.4	2.1	5.0	114	18.7	1.9	4.7
Women with employed husbands	38	27.8	2.9	6.5	24	26.3	2.5	7.0
Men with employed wives	31	16.7	1.7	4.1	38	15.7	1.7	3.7
Men with non-employed wives	29	14.8	1.4	3.9	51	17.3	1.7	4.3
Single mothers	14	28.8	2.8	7.4	21	30.2	2.8	8.5

*No difference between the means were significant at $p \leq .05$.

TABLE 17
MEAN WEEKLY HOURS ON HOME CHORES FOR MOTHERS AND NON-MOTHERS ON STANDARD TIME AND FLEXITIME*

	Standard Time			Flexitime		
Sample	*N*	*Mean Hours*	*SD*	*N*	*Mean Hours*	*SD*
Mothers	58	25.1	9.3	50	28.1	10.9
Non-mothers	24	17.2	7.3	26	21.8	7.5

*Differences not significant for mothers, significant for non-mothers at $p \leq .05$.

TABLE 18
RESPONDENTS' PERCENTAGE OF TIME ON HOME CHORES* (IN RELATION TO THEIR SPOUSES ON A WEEKLY BASIS)

	Standard Time			Flexitime		
Sample	*N*	*Mean*	*SD*	*N*	*Mean*	*SD*
Total sample	159	45.6	22.3	192	42.4	20.3
Parents	93	45.9	21.2	115	41.4	21.3
Women with employed husbands	38	63.5	17.0	26	70.6	18.2
Men with employed wives	31	42.1	11.1	38	41.0	11.2
Men with non-employed wives	24	23.1	10.0	50	26.1	9.4
Non-parents	60	44.4	23.1	72	43.8	19.2
Women with employed husbands	19	62.8	14.3	16	65.9	22.2
Men with employed wives	26	42.2	16.4	35	42.7	9.5
Men with non-employed wives	15	23.5	25.0	20	28.1	10.9

*No differences between the means were significant.

TABLE 19
RESPONDENTS' PERCENTAGE OF TIME WITH CHILDREN* (IN RELATION TO THEIR SPOUSES ON A WEEKLY BASIS)

	Standard Time			Flexitime		
Sample	*N*	*Mean*	*SD*	*N*	*Mean*	*SD*
Husband and wife families	88	50.5	20.6	106	42.2	20.1
Women with employed husbands	34	63.2	14.1	24	63.7	23.1
Men with employed wives	30	45.4	16.4	34	40.3	15.1
Men with non-employed wives	24	38.9	24.2	46	32.5	12.2

*No differences were significant at $p \leq .05$ level.

TABLE 20
MEAN DIFFERENCES* IN COMMUTING TIME TO AND FROM WORK FOR STANDARD TIME AND FLEXITIME RESPONDENTS (IN MINUTES)

Sample	*Standard Time*	*Flexitime*
To work		
Men with non-employed wives	48.8	50.0
Men with employed wives	43.4	44.3
Women with employed husbands	45.0	51.1
Single mothers	47.6	46.3
Men at high occupational level	45.2	45.4
Men at low occupational level	38.6	42.0
Women at high occupational level	37.4	39.4
Women at low occupational level	45.0	45.9
From work		
Men with non-employed wives	54.1	53.6
Men with employed wives	46.3	48.7
Women with employed husbands	47.2	54.3
Single mothers	53.9	48.9
Men at high occupational level	48.6	49.5
Men at low occupational level	44.7	45.5
Women at high occupational level	40.5	44.2
Women at low occupational level	48.5	49.4

*No differences between the groups were significant at $\leq .05$ level.

TABLE 21
COMPARISON OF TOTAL WEEKLY HOURS WORKED IN THE TWO AGENCIES

Sample	*Standard Time Mean*	*Flexitime Mean*	*Significance Level*
Men with non-employed wives	43.3	41.5	
Men with employed wives	45.6	42.3	*
Women with employed husbands	41.8	41.2	
Single mothers	42.1	39.6	
Men at high occupational level	44.9	41.9	*
Men at low occupational level	42.8	41.5	
Women at high occupational level	45.7	42.0	†
Women at low occupational level	40.7	40.5	

*Significant at $p \leq .01$.
†Significant at $p \leq .05$.

TABLE 22
RESPONDENT SATISFACTION WITH SPOUSE SHARE IN HOME CHORES, BY SEX AND BY AGENCY

	*Males**				*Females†*			
	Standard Time		*Flexitime*		*Standard Time*		*Flexitime*	
Spouse Share	*N*	*Percent*	*N*	*Percent*	*N*	*Percent*	*N*	*Percent*
Want spouse to spend:								
More time on chores		10.6	25	16.3	30	44.1	25	43.6
Less time on chores		39.4	50	33.5	5	7.4	5	9.4
Same time on chores		50.0	75	50.1	33	48.5	27	47.0
Total		100.0	150	100.0	68	100.0	57	100.0

*Chi square is 2.05136; significance is 0.3586.
†Chi square is 0.17467; significance is 0.9164.

TABLE 23
RESPONDENTS' ATTITUDES TOWARD THEIR SPOUSE'S SHARE IN CHILD REARING, BY SEX AND BY AGENCY

	*Males**				*Females†*			
	Standard Time		*Flexitime*		*Standard Time*		*Flexitime*	
Spouse Share	*N*	*Percent*	*N*	*Percent*	*N*	*Percent*	*N*	*Percent*
Want spouse to spend:								
More time with children	11	19.1	11	12.5	30	71.4	16	63.6
Less time with children	6	10.3	13	14.3	0	0.0	1	4.4
Same time with children	41	70.7	65	73.2	12	28.6	8	32.0
Total	58	100.0	89	100.0	42	100.0	25	100.0

*Chi square is 1.43429; significance is 0.4881.
†Chi square is 2.06012; significance is 0.3570.

Bibliography

A building triumph. 1931. *Manufacturer's Record*, Dec. 17, 36–41.

Abzug, Bella. 1975. Testimony. In U.S. Congress, House of Representatives, Committee on Post Office and Civil Service, Subcommittee on Manpower and Civil Service, *Alternate work schedules and part-time career opportunities in the federal government* (serial no. 94-53), pp. 2–8.

Adams, Richard N., and Jack N. Preiss. eds. 1960. *Human organization research: Field relations and techniques.* Homewood, Ill.: Dorsey Press.

Aldous, Joan. 1970. Strategies for developing family theory. *Journal of Marriage and the Family* 32: 250–57.

______. 1978. *Family, careers, and developmental changes in families.* New York: John Wiley.

Aldous, Joan, Marie W. Osmond, and Mary W. Hicks. 1979. Men's work and men's families. In *Contemporary theories about the family,* vol. 1, ed. Wesley R. Burr, Reuben Hill, F. Ivan Nye, and Ira L. Reiss. New York: Free Press.

Andrews, Frank M., James Morgan, and John Sonquist. 1967. *Multiple classification analysis: A report on a computer program for multiple regression using categorical predictors.* Ann Arbor, Mich.: Survey Research Center, Institute for Social Research, University of Michigan.

Andrews, Frank M., and Stephen B. Withey. 1976. *Social indicators of well-being: Americans' perception of life quality.* New York: Plenum Press.

Anshen, Ruth Nanda. 1959. The family: Its functions and destiny. Rev. ed. New York: Harper.

Ariès, Phillippe. 1962. *Centuries of childhood: A social history of family life.* New York: Knopf.

Arvey, Richard, and Ronald Gross. 1977. Satisfaction levels and correlates of satisfaction in the homemaker job. *Journal of Vocational Behavior* 10: 13–24.

Bahr, Stephen. 1974. Effects on power and division of labor in the family. In *Working Mothers*, ed. L. Hoffman and F. Ivan Nye. San Francisco: Jossey Bass.

Bailyn, Lotte. 1970. Career and family orientations of husbands and wives in relation to marital happiness. *Human Relations* 23: 97–113.

______. 1973. Family constraints on women's work. *Annals of the New York Academy of Science* 208: 82–90.

______. 1977. Research as a cognitive process: Implications for data analysis. *Quality and Quantity* 11: 97–117.

Balswick, J. O., and C. W. Peek. 1971. The inexpressive male: A tragedy of American society. *The Family Coordinator* 20: 363–68.

Bane, Mary Jo. 1976. *Here to stay: American families in the twentieth century.* New York: Basic Books.

Barker, R. G. 1965. Explorations in ecological psychology. *American Psychologist* 20: 1–14.

______. 1968. *Ecological psychology: Concepts and methods for studying the environment of human behavior.* Stanford, Calif.: Stanford University Press.

Bebbington, A. C. 1973. The functions of stress in the establishment of the dual-career family. *Journal of Marriage and the Family* 35: 530–37.

Becker, Gary. 1965. A theory of the allocations of time. *The Economic Journal* 75: 493–517.

Becker, Howard S., Blanche Geer, David Riesman, and Robert S. Weiss. 1968. *Institutions and the person: Papers presented to Everett C. Hughes.* Chicago: Aldine.

Bennett, Sheila, and Glen H. Elder. 1979. Women's work in the family economy: A study of depression hardship in women's lives. *Journal of Family History* 4 (2): 153–76.

Berg, Ivar. 1971. *Education and jobs: The great training robbery.* Boston: Beacon Press.

Berheide, Catherine W., and Richard A. Berk. 1976. Household work in the suburbs: The job and its participants. *Pacific Sociological Review* 19: 491–518.

Berheide, Catherine W., and Sarah F. Berk. 1977. Going backstage—gaining access to observe household work. *Sociology of Work and Occupations* 4 (1): 27–48.

Berkove, Gail Feldman. 1979. Perceptions of husband support by returning women students. *The Family Coordinator* 28 (4): 451–57.

Bernard, Jessie Shirley. 1942. *American family behavior.* New York: Harper.

______. 1949. *American community behavior.* Hinsdale, Ill.: Dryden Press.

______. 1972. *The future of marriage.* New York: World.

______. 1974. *The future of motherhood.* New York: Dial Press.

Best, Fred, and Barry Stern. 1977. Education, work, and leisure: Must they come in that order? *Monthly Labor Review* 100 (7): 31–34.

Biller, H. B. 1976. The father and personality development: Paternal deprivation and sex-role development. In *The role of the father in child development*, ed. M.E. Lamb. New York: John Wiley.

Blau, Peter M., and Otis Dudley Duncan. 1967. *The American occupational structure.* New York: John Wiley.

Blood, Robert O., and Donald M. Wolfe. 1960. *Husbands and Wives.* Glencoe, Ill.: Free Press.

Bombeck, Erma. 1978. *If life is just a bowl of cherries, what am I doing in the pits?* New York: Fawcett.

Booth, Alan. 1977. Wife's employment and husband's stress: A replication and refutation, *Journal of Marriage and the Family* 39: 645–50.

______. 1979. Does wives' employment cause stress for husbands? *The Family Coordinator* 28 (4): 145–449.

Borow, Henry. 1967. *Man in the world of work.* Boston: Houghton Mifflin.

Bowlby, J. 1951. *Maternal care and mental health.* Geneva: World Health Organization.

______. 1971. *Attachment.* Attachment and loss, vol. 1. London: Hogarth Press, 1969; Harmondsworth, England: Penguin.

______. 1972. *Child care and the growth of love.* Harmondsworth, England: Penguin.

______. 1973. *Separation.* Attachment and loss, vol. 2. London: Hogarth Press.

Brim, O. G. 1975. Macro-structural influences in child development and the need for childhood social indicators. *American Journal of Orthopsychiatry* 45: 516–24.

Broderick, C. B., ed. 1971. *A decade of family research and action, 1960–69.* National Council on Family Relations.

Bronfenbrenner, Urie. 1977a. Testimony. In U.S., Congress, House of Representatives, Committee on Post Office and Civil Service, Subcommittee on Employee Ethics and Utilization, *Part-time employment and flexible work hours* (serial no. 95-28).

_____. 1977b. Toward an experimental ecology of human development. *American Psychologist* 32 (7): 513–31.
_____. 1979. *Ecology of human development.* Cambridge, Mass.: Harvard University Press.
Bronfenbrenner, Urie, Amy Avgar, and Charles R. Henderson, Jr. 1977. An analysis of family stresses and supports: A progress report on a pilot study. Mimeograph: draft of a paper from research program on the comparative ecology of human development, Cornell University.
Bronfenbrenner, Urie, and Moncrieff Cochran. 1976. Conceptual framework: Categories for assessing parental stresses and supports. Mimeograph for Cornell six-nation study, Cornell University.
Bueche, Nancy A., and Ramona Marotz-Baden. 1978. Decision-making in two types of dual-employed families. Paper presented at the National Council on Family Relations, Philadelphia.
Burchard, W. 1954. Role conflicts of military chaplains. *American Sociological Review* 19: 528–35.
Bureau of the Census. 1976. Census Special Studies, no. 58, April.
Bureau of Labor Statistics. 1975. *Travel-to-work survey of the annual housing survey.* Washington, D.C.: U.S. Department of Labor.
_____. 1977. *U.S. working women: A databook.* Washington, D.C., U.S. Department of Labor.
_____. 1978. *Marital and family characteristics of the labor force,* March 78–638 . Washington, D.C.: U.S. Department of Labor.
Burke, Yvonne B. 1975. Testimony. In U.S., Congress, House of Representatives, Committee on Post Office and Civil Service, Subcommittee on Manpower and Civil Service Hearings, *Alternate work schedules and part-time career opportunities in the federal government* (serial no. 94-53), pp. 14–19.
_____. 1977. Testimony. In U.S., Congress, House of Representatives, Committee on Post Office and Civil Service, Subcommittee on Employee Ethics and Utilization Hearings, *Part-time employment and flexible work hours* (serial no. 95-28), pp. 49–54.
_____. 1978. Testimony. In U.S., Congress, Senate, Committee on Governmental Affairs, *Flexitime and part-time legislation* (serial no. .32-422 0), pp. 16–20.
Burr, Wesley R. 1973. *Theory construction and the sociology of the family.* New York: John Wiley.

Burr, Wesley R., Reuben Hill, F. Ivan Nye, and Ira L. Reiss, eds. 1979. *Contemporary theories about the family*. New York: Free Press.

Business Week, 1972. Europe likes flexi-time work, Oct. 7, pp. 80–82.

______. 1973. Flexible hours, March 10, p. 49.

Campbell, Angus, Phillip E. Converse, and Willard L. Rodgers. 1976. *The quality of American life: Perceptions, evaluations, and satisfactions*. New York: Russell Sage Foundation.

Campbell, Donald T., and Julian C. Stanley. 1963. *Experimental and quasi-experimental designs for research*. Chicago: Rand McNally.

Campbell, R. E. 1969. Vocational ecology: A perspective for the study of careers. *The Counseling Psychologist* 1: 20–23.

Caplow, Theodore. 1954. *The sociology of work*. Minneapolis: University of Minnesota Press.

Carter, Jimmy. 1978. Statement by the President on signing H.R. 7814. Press release, Office of the White House Press Secretary, Sept. 29.

Causey, Mike. 1976. The federal diary: More flexitime due. *The Washington Post*, Dec. 21.

______. 1978. Part-time job bill enacted. *The Washington Post*, Sept. 28.

Chafe, William H. 1976. Looking backward in order to look forward: Women, work, and social values in America. In *Women and the American economy*, ed. Juanita M. Kreps. Englewood Cliffs, N.J.: Prentice-Hall.

Cherlin, Andrew. 1978. *Employment, income, and family life: The case of marital dissolution in women's changing roles at home and on the job*. Special report no. 26, National Commission for Manpower Policy, Washington, D.C., Sept. 1978.

Christensen, H. T. 1964. Development of the family field of study. In *Handbook of marriage and the family*, ed. H. T. Christensen. Chicago: Rand McNally.

Clark, Robert A., F. Ivan Nye, and Viktor Gecas. 1978. Work involvement and marital role performance. *Journal of Marriage and the Family* 40 (1): 9–22.

Clarke-Stewart, Alison. 1977. *Child care in the family: A review of research and some propositions for policy*. New York: Academic Press.

Clayre, Alasdair. 1974. *Work and play: Ideas and experience of work and leisure*. New York: Harper and Row.

Cochran, Moncrieff, and Phyllis Moen. 1977. Work, families, and the developing child: A systematic perspective. Mimeograph prepared by Cornell University for the Committee on Child Development Research and Public Policy of the National Research Council, National Academy of Sciences, May.

Cohen, Allan R. 1978. Personal communication regarding study of effects of alternate work schedules at the John Hancock Company, Boston.

Cohen, Allan R., and Herman Gadon. 1978. *Alternative work schedules and the adult life cycle: Individual and organization uses of time.* Reading, Mass.: Addison-Wesley.

Cohen, Lynne Revo. 1978. Testimony. In U.S., Congress, Senate, Committee on Governmental Affairs, *Flexitime and part-time legislation* (serial no. .32-422 O), pp. 82–83.

Coiner, Maryrose Carew. 1978. Employment and mothers' emotional states: A psychological study of women re-entering the work force. Ph.D. diss., Yale University.

Colletta, Nancy Donohoe. 1978. Divorced mothers at two income levels: Their support systems and child-rearing practices. Ph.D. diss., Cornell University.

Communications Workers of America, AFL-CIO. 1978. Six-union delegation completes study of alternate work patterns in Europe. Newsletter of CWA, March 20.

Control Data Corporation. N.d. General information on flexible work hours. Mimeograph.

Cook, A. 1975. *The working mother.* Ithaca, N.Y.: New York State School of Industrial and Labor Relations, Cornell University.

Coser, Lewis A. 1974. *Greedy institutions: Patterns of undivided commitment.* New York: Collier MacMillan.

Cowley, Thomas. 1976. Flexitime for increased productivity. Mimeograph. Pay Policy Division, Office of Personnel Management, May.

______. 1979. Personal communication. Office of Personnel Management, Pay and Leave Section.

Cox, M. 1974. The effects of father absence and maternal employment on the development of children. Ph.D. diss., University of Virginia.

Crites, J. P. 1969. *Vocational psychology.* New York: McGraw-Hill.

Cronbach, L., and P. Meehl. 1955. Construct validity of psychological tests. *Psychological Bulletin* 52: 281–302.

Croog, Sydney H. 1970. The family as a source of stress. In *Social Stress,* ed. Sol Levine and Norman A. Scotch. Chicago: Aldine.

Crowne, D. P., and D. Marlowe. 1964. *The approval motive.* New York: John Wiley.

Current Population Reports. 1977. *Marital status and living arrangements: March 1976.* Series P-20, no. 306. Washington, D.C.: U.S. Department of Commerce, Bureau of the Census.

Dahlstrom, E., and R. Liljestrom, eds. 1971. *The changing roles of men and women.* Boston: Beacon Press.

Datan, N., ed. 1975. *Life-span development psychology: Normative life crises.* New York: Academic Press.

De Frain, J. D. 1974. A father's guide to parent guides: Review and assessment of the paternal role as conceived in the popular literature. Paper delivered at the National Council on Family Relations/ American Association of Marriage and Family Counselors Annual Meeting.

De Grazia, Sebastian. 1962. *Of time, work, and leisure.* New York: Twentieth Century Fund.

deMausse, Lloyd. 1974. The evolution of childhood. In *The history of childhood,* ed. Lloyd deMausse. New York: Harper and Row.

Demos, John. 1974. The American family in past time. *American Scholar* 43: 422–46.

Denzin, Norman K. 1978. *Sociological methods: A sourcebook.* 2d ed. Chicago: Aldine.

Dexter, Lewis Anthony. 1970. *Elite and specialized interviewing.* Evanston, Ill. Northwestern University Press.

Dise, Christine. 1977. Personal communication. Work Force Analysis and Statistics Division, Civil Service Commission.

Dizard, J. 1968. *Social change in the family.* Chicago: University of Chicago Press.

Dohrenwend, B. 1973. Life events as stressors: A methodological inquiry. *Journal of Health and Social Behavior* 14 (2): 167–75.

Drabble, Margaret. 1971. *Eye of the needle.* London: Penguin.

______. 1976. *Realms of gold.* London: Penguin.

Duckles, Margaret. 1977. Wives and mothers working. Ph.D. diss., Wright Institute Graduate School.

Dullea, Georgia. 1978. Fast changes in society traced to the rise of working women. *The New York Times,* Nov. 29.

Dyer, William G. 1964. Family reaction to the father's job. In *Blue-collar world: Studies of the American worker,* ed. A. Shostak and W. Gomberg. Englewood Cliffs, N.J.: Prentice-Hall.

Ehrenreich, Barbara. 1979. Is success dangerous to your health? The myths—and facts—about women and stress. *Ms. Magazine* 7 (11): 51–54, 97–99.

Elbing, Alvar O., Herman Gadon, and John R. M. Gordon. 1974. Flexible working hours: It's about time. *Harvard Business Review* 52: 18–20.

Elder, Glenn H. 1979. Comments at theory construction workshop, National Council on Family Relations, Annual Meeting, Boston, Aug.

Epstein, Cynthia Fuchs. 1970. *Woman's place: Options and limits in professional careers.* Berkeley: University of California Press.

Erikson, Eric. 1963. *Childhood and society.* Rev. ed. New York: Norton.

Etaugh, Claire. 1974. Effects of maternal employment on children: A review of recent research. *Merrill-Palmer Quarterly* 20: 71–98.

Etzioni, Amitai, and Carol O. Atkinson. 1969. *Sociological implications of alternative income transfer systems.* New York: Bureau of Social Science Research, Columbia University.

Evans, A. 1973. *Flexibility in working life: Opportunities for individual choice.* Paris: Organization for Economic Cooperation and Development.

Eyde, Lorraine D. 1975. Testimony. In U.S., Congress, House of Representatives, Committee on Post Office and Civil Service, Subcommittee on Manpower and Civil Service Hearings, *Alternate work schedules and part-time career opportunities in the federal government* (serial no. 94-53), pp. 151–74.

Family Impact Seminar. 1978a. *Toward an inventory of federal programs with direct impact on families.* Washington, D.C.: Institute for Educational Leadership, The George Washington University, Feb.

______. 1978b. *Interim Report.* Washington, D.C.: Institute for Educational Leadership, The George Washington University, April.

______. 1979. *Field coordinator's guide: Family impact field project.* Washington, D.C.: Institute for Educational Leadership, The George Washington University, Oct.

Farr, Louise. 1978. Personal communication. Director of admissions, Georgetown University Law School.

Farrell, W. 1974. *The liberated man: Beyond masculinity, freeing men, and their relationship with women.* New York: Random House.

Feigen-Fasteau, Marc. 1974. *The male machine.* New York: McGraw-Hill.

Fein, R. 1978. Research on fathering: Social policy and an emergent perspective. *Journal of Social Issues* 34 (1): 122–35.

Feldman, H., and M. Feldman. 1975. The effects of father absence on adolescence. Paper delivered at the National Council of Family Relations, Salt Lake City, Aug.

Fellows, Lawrence. 1971. German workers setting own hours. *The New York Times*, July 12.

Ferree, Myra. 1976. Working class jobs: Housework and paid work as sources of satisfaction. *Social Problems* 23: 431–41.

Fidell, Linda. 1976. Employment status, role dissatisfaction, and the housewife syndrome. Mimeograph. Department of Psychology, California State University at Northridge.

Finegan, James. 1977. Testimony. In U.S., Congress, House of Representatives, Committee on Post Office and Civil Service, Subcommittee on Employee Ethics and Utilization Hearings, *Part-time employment and flexible work hours* (serial no. 95-28), pp. 155–65.

Fiss, Barbara. 1977. Testimony. In U.S., Congress, House of Representatives, Committee on Post Office and Civil Service, Subcommittee on Employee Ethics and Utilization Hearings, *Part-time employment and flexible work hours* (serial no. 95-28), pp. 12–23.

Fiss, Barbara, and Thomas Cowley. 1977. Personal communication. Office of Personnel Management, Pay and Leave Section.

Flynn, Judy. 1977. The implications of flexitime: Particularly for women. Paper prepared for the National Organization for Women, Jan. 31.

Fogarty, Michael P., Rhona Rapoport, and Robert N. Rapoport. 1971. *Sex, career, and family*. London: George Allen and Unwin.

Frankenhaeser, Marianne, and Bertil Gardell. 1976. Underload and overload in working life: Outline of a multidisciplinary approach. *Journal of Human Stress*, pp. 35–46.

Freud, S. 1905/43. *Complete works of Sigmund Freud*, vol. 7. (Standard ed.) London: Hogarth Press.

Friedan, Betty. 1963. *The feminine mystique*. New York: Norton.

Furstenberg, Frank F., Jr. 1974. Work experience and family life. In *Work and the quality of life: Resource papers for work in America*, ed. James O'Toole. Cambridge, Mass.: MIT Press.

Gans, Herbert J. 1962. *The urban villagers*. New York: Free Press.

Gavan, Bill. 1977. Personal Communication. Office of Management and Budget.

Gertsl, Joel E. 1961. Leisure, taste, and occupation milieu. *Social Problems* 9: 56–68.

Gerzon, Mark. 1970. *A childhood for every child: The politics of parenthood.* New York: E. P. Dutton.

Gilman, Charlotte Perkins. 1903. The home: Its work and influence. *Female Liberation*, ed. Roberta Salper. New York: Knopf, 1972.

Ginzberg, Eli. 1975. *The manpower connection.* Cambridge, Mass.: Harvard University Press.

Giraldo, Z. I., and J. M. Weatherford. 1978. *Life cycle and the American family: Current trends and policy implications.* Durham, N.C.: Institute of Policy Sciences and Public Affairs, Duke University.

Gladstone, Leslie. 1977. Testimony. In U.S., Congress, House of Representatives, Committee on Post Office and Civil Service, Subcommittee on Employee Ethics and Utilization, *Part-time employment and flexible work hours* (serial no. 95-28), p. 33.

Glazer-Malbin, Nona. 1976. Review essay: Housework. *Journal of Women in Culture and Society* 1 (4): 905–22.

Glick, Paul C. 1977. Updating the life cycle of the family. *Journal of Marriage and the Family* 39: 3–15.

Glickman, Albert S., and Zeniz H. Brown. 1973. Changing schedules of work: Patterns and implications. Springfield, Va.: National Technical Information Service.

Golden, M. Patricia, ed. 1976. *The research experience.* Itasca, Ill.: F. E. Peacock.

Golembiewski, Robert T., Rick Hilles, and Munro S. Kagno. 1974. Longitudinal study of flexi-time effects: Some consequences of an OD structural intervention. *Journal of Applied Behavioral Science* 10 (4): 503–32.

Golembiewski, Robert T., and Carl W. Proehl, Jr. 1980. Public sector applications of flexible workhours: A review of available experience. *Public Administration Review* 1 (1): 72–85.

Golembiewski, Robert T., Samuel Yeager, and Rick Hilles. 1975. Factor analysis of some flexitime effects: Attitudinal and behavioral consequences of a structural intervention. *Academy of Management Journal* 18 (3): 500–509.

Goode, William J. 1959. Horizons in family theory. In *Sociology today: Problems and prospects*, ed. Robert King Merton, Leonard Brown, and Leonard S. Cuttrell, Jr. New York: Basic Books.

______. 1960. A theory of role strain. *American Sociological Review* 25 (4): 483–96.

Greenwald, Carol. 1978. Testimony. In U.S., Congress, Senate, Committee on Governmental Affairs, *Flexitime and part-time legislation* (serial no. .32-422 O), pp. 38–39, 45–46.

Greven, Philip. 1970. *Four generations: Population, land, and family in colonial Andover, Massachusetts.* New York: Cornell University Press.

Gronseth, Erik. 1971. The husband provider role: A critical appraisal. In *Family issues of employed women in Europe and America*, ed. Andrée Michel. Leiden: E. J. Brill.

______. 1978. Work sharing: A Norwegian example. In *Working couples*, ed. Robert N. Rapoport, Rhona Rapoport, and Janice Bumstead. New York: Harper Colophon Books.

Gross, N., W. S. Mason, and A. W. McEarchern. 1958. *Explorations in role analysis: Studies of the school superintendency role.* New York: John Wiley.

Grossman, Allyson Sherman. 1979. *Employment in perspective: Working women.* Report 565, Washington, D.C.: Bureau of Labor Statistics, U.S. Department of Labor.

Gutman, Herbert G. 1976. *Work, culture, and society in industrializing America.* New York: Knopf.

Guttentag, M., and S. Salasin. 1975. Women, men and mental health. In *Women and men: Changing roles and perceptions*, ed. L. A. Cater and A. F. Scott. Stanford, Calif.: Aspen Institute.

Haavio-Mannila, E. 1971. Satisfaction with family, work, leisure, and life among men and women. *Human Relations* 24: 585–601.

Haldi Associates. 1977. Summary report of findings of a comprehensive technology assessment of alternative work schedules. New York: Haldi Associates.

Haley, Jay. 1976. *Problem solving therapy.* San Francisco: Jossey-Bass.

Hamburger, Martin, and Howard Hess. 1970. Work performance and emotional disorder. In *Mental health and work organizations*, ed. A. A. McLean. Chicago: Rand McNally.

Hansen, Donald A., and Vicky A. Johnson. 1979. Rethinking family stress theory: Definitional aspects. In *Contemporary theories about the family*, vol. 1, ed. Wesley R. Burr, Reuben Hill, F. Ivan Nye, and Ira L. Reiss. New York: Free Press.

Harden, Blaine. 1979. The painful questions: Stress of work v. family decisions is area's major mental health problem. *The Washington Post*, April 9.

Hareven, Tamara K. 1975. Family time and industrial time: Family and work in a planned corporation town, 1900–24. *Journal of Urban History* 1 (3): 365–89.

______. 1976. Modernization and family history: Perspectives on social change. *Signs: Journal of Women in Culture and Society* 2 (1): 190–206.

Hareven, Tamara K., and Randolph Langenbach. 1978. *Amoskeag: Life and work in an American factory-city.* New York: Pantheon.

Hartup, W. W. 1978. Perspectives on child and family interaction: Past, present, and future. In *Child influences on marital and family interaction: A life-span perspective,* ed. R. M. Lerner and G. B. Spanier. New York: Academic Press.

Hayghe, Howard. 1976. Families and the rise of working women: An overview. *Monthly Labor Review* 99 (5): 12–19.

Hedges, Janice Neipert. 1977. Flexible schedules: Problems and issues. *Monthly Labor Review* 100 (2): 62–65.

Hedges, Janice Neipert, and Jeanne Barnett. 1972. Working women and the division of household tasks. *Monthly Labor Review* 95 (4): 9–14.

Hendrickson, Gladys. 1975. Testimony. In U.S., Congress, House of Representatives, Committee on Post Office and Civil Service, Subcommittee on Manpower and Civil Service Hearings, *Alternative work schedules and part-time career opportunities in the federal government* (serial no. 95-28), pp. 67–73.

Henry, Jules. 1965. *Pathways to madness.* New York: Random House.

Herman, Jeanne, and Karen Gyllstrom. 1977. Working men and women: Inter- and intra-role conflict. *Psychology of Women Quarterly* 1: 319–33.

Hetherington, E. M. 1972. Effects of paternal absence on personality development in adolescent daughters. *Developmental Psychology* 7: 313–26.

Hetherington, E. M., et al. 1977. The development of children of mother headed families. Paper delivered at the Families in Contemporary America conference at The George Washington University, June 11.

Hicks, Nancy. 1976. Family stress called a menace to health. *The New York Times,* Oct. 20.

Hilgert, R., and J. Hundley. 1975. Supervision: The weak link in flexible work scheduling. *The Personnel Administrator* 20 (1): 24–26.

Hill, Reuben, and Donald A. Hansen. 1960. The identification of conceptual frameworks utilized in family study. *Marriage and Family Living* 22: 299–311.

Hodgson, James W., and Robert A. Lewis. 1979. Pilgrim's progress III: A trend analysis of family theory and methodology. *Family Process* 18 (2): 163–74.

Hofferth, Sandra L., and Kristin A. Moore. 1979. Women's employment and marriage. In *The subtle revolution: Women at work*, ed. Ralph E. Smith. Washington, D.C.: The Urban Institute.

Hoffman, Lois Wladis. 1960. Effects of the employment of mothers on parental power relations and the division of household tasks. *Marriage and Family Living* 22: 27–35.

______. 1974. Effects of maternal employment on the child: A review of the research. *Developmental Psychology* 10: 204–208.

Hoffman, Lois Wladis, and F. Ivan Nye. 1974. *Working mother.* San Francisco: Jossey-Bass.

Holmes, T. S., and R. H. Rahe. 1967. The social readjustment rating scale. *Journal of Psychosomatic Research* 11: 213–18.

Holmstrom, Linda Lytle. 1972. *The two-career family.* Cambridge, Mass.: Schenkman.

Hood, Jane C. 1977. Work-scheduling, work-involvement, and parental role-sharing in sixteen two-job families. Paper delivered to the Task Force on the "Impact of Business and Industry on the Family," National Council of Family Relations, April 11.

Hood, Jane C., and Susan Golden. 1979. Beating time/making time: The impact of work scheduling on men's family roles. *The Family Coordinator* 28 (4): 575–82.

Hooper, Judith Oakey. 1979. My wife, the student. *The Family Coordinator* 28 (4): 459–64.

Howrigan, Gail. 1973. Effects of working mothers on children. Center for the Study of Public Policy, Cambridge, Mass., Aug.

Hunt, David. 1970. *Parents and children in history: The psychology of family life in early modern France.* New York: Basic Books.

Hunt, J. G., and L. L. Hunt. 1977. Dilemmas and contradictions of status: The case of the dual-career family. *Social Problems* 24 (4): 407–16.

Ilfeld, Frederic W. 1977. Current social stressors and symptoms of depression. *American Journal of Psychiatry* 134 (2): 161–66.

Illick, Joseph E. 1974. Child-rearing in seventeenth-century England and America. In *The history of childhood*, ed. Lloyd deMausse. New York: Harper and Row.

International Labour Organization. 1973. *Part-time employment: An in-*

ternational survey. Geneva, Switzerland: International Labour Organization.

Jacobs, Glenn, ed. 1970. *The participant observer.* New York: George Braziller.

Johnson, Beverly L., and Howard Hayghe. 1977. Labor force participation of married women, March 1976. *Monthly Labor Review* 100 (6): 32–36.

Johnson, Frank A., and Colleen L. Johnson. 1976. Role strain in high-commitment career women. *Journal of the American Academy of Psychoanalysis* 4 (1): 13–36.

______. 1977. Attitudes towards parenting in dual-career families. *American Journal of Psychiatry* 134 (4): 391–95.

Johnson, John M. 1975. *Doing field research.* New York: Free Press.

______. 1977. Symposium on qualitative methods: Editorial introduction. *Urban Life* 6 (3): 329–32.

Kagan, Jerome, Richard B. Kearsley, and Philip R. Zelazo. 1978. *Infancy: Its place in human development.* Cambridge, Mass.: Harvard University Press.

Kamerman, Sheila B. 1976. Developing a family impact statement. Mimeograph. An occasional paper from the Foundation for Child Development, New York, May.

______. 1978a. Testimony. In U.S., Congress, House of Representatives, the Select Committee on Population, May 25.

______. 1978b. Work and family in industrialized societies. Mimeograph. Cross National Studies of Social Services and Family Policy, Columbia University, Sept. 21.

______. 1980. *Parenting in an unresponsive society: Managing work and family life.* New York: Free Press.

Kamerman, Sheila B., and Alfred J. Kahn. 1979. The day care debate: A wider view. *The Public Interest* (Winter, no. 54): 76–93.

______. 1980. *Child care, family benefits, and working parents.* New York: Columbia University Press.

Kanter, Rosabeth Moss. 1977. *Work and family in the United States: A critical review and agenda for research and policy.* New York: Russell Sage Foundation.

Kantor, David, and William Lehr. 1975. *Inside the family.* San Francisco: Jossey-Bass.

Keniston, Kenneth, and the Carnegie Council on Children. 1977. *All our children: The American family under pressure.* New York: Harcourt Brace Jovanovich.

Keppler, Bernard. 1979. Personal communication. Interflex Corporation.

Kerlinger, Fred N. 1973. *Foundations of behavioral research.* New York: Holt, Rinehart, and Winston.

Kerr, Jean. 1960. *Please don't eat the daisies.* New York: Fawcett.

Kish, Leslie. 1965. *The survey sample.* New York: John Wiley.

Kistler, Frederick. 1977. Testimony. In U.S., Congress, House of Representatives, Committee on Post Office and Civil Service, Subcommittee on Employee Ethics and Utilization Hearings, *Part-time employment and flexible work hours* (serial no. 95-28), pp. 12–23.

Kohn, Melvin L. 1977. *Class and conformity: A study in values.* 2nd ed. Chicago: The University of Chicago Press.

Kohn, Melvin L., and Carmi Schooler. 1973. Occupational experience and psychological functioning: An assessment of reciprocal effects. *American Sociological Review* 38: 97–118.

Komarovsky, Mirra. 1964. *Blue-collar marriage.* New York: Random House.

______. 1977. *Dilemmas of masculinity.* New York: Norton.

Kreps, Juanita M. 1968. *Lifetime allocations of work and leisure.* Washington, D.C.: Office of Research Statistics, Social Security Administration, HEW.

______, ed. 1976. *Women and the American economy.* Englewood Cliffs, N.J.: Prentice-Hall.

Kronholz, June. 1978. Women at work: Management practices change to reflect role of women employees. *Wall Street Journal*, Sept. 13.

Kuhn, Thomas S. 1970. *The structure of scientific revolutions.* Chicago: University of Chicago Press.

Kuhne, Robert J., and Courtney O. Blair. 1978. Flexitime. *Business Horizons* 21 (2): 29–44.

La Bier, Douglas. 1980. Uncle Sam's working wounded. *The Washington Post Magazine*, Feb. 17.

Lamb, M. E., ed. 1976. *The role of the father in child development.* New York: John Wiley.

Lamm, Dottie. 1977. Testimony. In U.S., Congress, House of Representatives, Committee on Post Office and Civil Service, Subcommittee on Employee Ethics and Utilization Hearings, *Part-time employment and flexible work hours* (serial no. 95-28), pp. 107, 110.

Langholz, B. 1972. Variable working hours in Germany. *Journal of Systems Management* 23: 30–33.

Langner, G. S., and S. T. Michaels. 1963. *Life stresses and mental health.* New York: Free Press.

Lasch, Christopher. 1977. *Haven in a heartless world: The family besieged.* New York: Basic Books.

Laslett, Barbara. 1973. The family as a public and private institution: An historical perspective. *Journal of Marriage and the Family* 35 (3): 480–92.

Laslett, Peter. 1977. Characteristics of the western family considered over time. *Journal of Family History* 2 (2): 89–115.

Lauer, R. H., and J. C. Lauer. 1976. The experience of change: Tempo and stress. In *Social change: Explorations, diagnoses, and conjectures,* ed. G. K. Zollschan and W. Hirsch. Cambridge, Mass.: Schenkman.

Lazar, Irving, V. R. Hubbell, H. Murray, M. Rosche, and J. Royce. 1977. *The persistence of preschool effects final report.* Ithaca, N.Y.: Community Services Education Laboratory.

Legge, K. 1974. Flexible working hours: Panacea or placebo? *Management Decision* 12 (5): 264–79.

Lein, Laura, Jan Lennon, Maureen Durham, Gail Howrigan, Michael Pratt, Ronald Thomas, and Heather Weiss. 1974. *Work and family life.* Final report to the National Institute of Education. Cambridge, Mass.: Center for the Study of Public Policy.

Levine, James. 1976. *Who will raise the children? New options for fathers (and mothers).* Philadelphia: Lippincott.

Levinson, Harry. 1964. *Executive stress.* New York: Harper and Row.

______. 1971. Various approaches to understanding man at work. *Architectural Environmental Health* 22: 612–18.

Levitan, Sar A., G. L. Magnum, and Ray Marshall. 1972. *Human resources and labor markets.* New York: Harper and Row.

Lewin, K. 1935. *A dynamic theory of personality.* New York: McGraw-Hill.

______. 1936. *Problems of topological psychology.* New York: McGraw-Hill.

______. 1948. *Resolving social conflict.* New York: Harper.

______. 1951. *Field theory in social science.* New York: Harper.

Lewis, J. M. 1976. *No single thread: Psychological health in the family systems.* New York: Bruner-Maisel.

Liljestrom, Rita. 1978. Integration of family policy and labor market policy in Sweden. *Social Change in Sweden* (Dec. no. 9): 1–9.

Linder, Steffon B. 1970. *The harried leisure class.* New York: Columbia University Press.

Lloyd, Cynthia, ed. 1975. *Sex, discrimination, and the division of labor.* New York: Columbia University Press.

Lopata, Helena S. 1971. *Occupation: Housewife.* New York: Oxford University Press.

Lynn, D. B. 1974. *The father: His role in child development.* Belmont, Calif.: Brooks/Cole.

Maccoby, Eleanor E. 1958. Effects upon children of their mother's outside employment. In *Work in the lives of married women,* National Manpower Council. New York: Columbia University Press.

Macke, Anne S., and Paula M. Haudis. 1977. Sex-role attitudes and employment among women: A dynamic model of change and continuity. Paper delivered at the Secretary of Labor's Invitational Conference on the National Longitudinal Surveys of Mature Women, Washington, D.C., Jan. 26.

Magnusson, Monica, and Carina Nilsson. 1979. *Att arbeta pa obekväm arbetstid: Vad säger fovskningen om konsekvenser för människan och behov av förbättringar beträffande arbetstidens förläggning?* (report no. 91-518-1293-2). Stockholm, Sweden: Preiser in cooperation with the Swedish Work Environment Fund.

Mainardi, P. 1972. The politics of housework. In *Woman in a man-made world,* ed. Nona Glazer-Malbin and Helen Y. Waehrer. Chicago: Rand McNally.

Maklan, David M. 1977. How blue collar workers on four day workweeks use their time. *Monthly Labor Review* 100 (8).

Marks, Stephen. 1977. Multiple roles and role strain: Some notes on human energy, time, and commitment. *American Sociological Review* 42: 921–36.

Marsh, Fay. 1977. Testimony. In U.S., Congress, House of Representatives, Committee on Post Office and Civil Service, Subcommittee on Employee Ethics and Utilization Hearings, *Part-time employment and flexible work hours* (serial no. 95-28), pp. 101–106.

Martin, Virginia Hilder. 1974. Recruiting women workers through flexible hours. *S.A.M. Advanced Management Journal,* (July), pp. 46–53.

Martin, Virginia Hilder, and Jo Hartley. 1975. *Hours of work when workers can choose.* Washington, D.C.: Research Project of the Business and Professional Women's Foundation.

Maslow, A. H. 1954. *Motivation and personality.* New York: Harper and Row.

______. 1962. *Toward a psychology of being*. Princeton, N.J.: Van Nostrand.

Mason, K. O., and L. L. Bumpass. 1975. U.S. women's sex-role ideology, 1970. *American Journal of Sociology* 80 (5): 1212–19.

Mattessich, P. 1977. *Family impact analysis*. St. Paul, Minn.: State Planning Agency, Spring.

McCall, George J., and J. L. Simmons, eds. 1969. *Issues in participant observation: A text and reader.* Reading, Mass.: Addison-Wesley.

McGrath, J. E. 1970. *Social and psychological factors in stress*. New York: Holt, Rinehart, and Winston.

McHugh, Ed. 1980. Personal communication. Office of Personnel Management, Office of Policy Analysis and Development.

McLaughlin, Virginia Yans. 1971. Patterns of work and family organization: Buffalo's Italians. *Journal of Interdisciplinary History* 2 (2): 299–314.

Mechanic, D. 1962. *Students under stress*. Glencoe, Ill.: Free Press.

______. 1974. Social structure and personal adaptation: Some neglected dimensions. In *Coping and adaptation*, ed. G. V. Coelho, D. Haniburg, and J. F. Adams. New York: Basic Books.

______. 1975. Some problems in the measurement of stress and social readjustment. *Journal of Human Stress* 1: 43–48.

Meier, Gretl S. 1978. *Job sharing*. Kalamazoo, Mich.: W. E. UpJohn Institute for Employment Research.

Meissner, Martin, Elizabeth Humphries, Scott Meis, and William Scheu. 1975. No exit for wives: Sexual division of labor and the cumulation of household demands. *Review of Canadian Sociology and Anthropology* 12: 424–39.

Melia, Jinx. 1975. Testimony. In U.S., Congress, House of Representatives, Committee on Post Office and Civil Service, Subcommittee on Manpower and Civil Service Hearings, *Alternate work schedules and part-time career opportunities* (serial no. 94-53), pp. 73–77.

Merton, Robert King. 1945. Sociological theory. *American Journal of Sociology* 50: 462–73.

Michel, Andrée, ed. 1971. *Family issues of employed women in Europe and America*. Leiden: E. J. Brill.

Ministry of Health and Social Affairs, International Secretariat. 1977. *Parental insurance in Sweden—some data*. Stockholm, Sweden.

Minturn, Leigh, and William W. Lambert. 1964. *Mothers of six cultures: Antecedents of child rearing*. New York: John Wiley.

Minuchin, Salvador. 1975. *Families and family therapy.* Cambridge, Mass.: Harvard University Press.

Montagna, Paul. 1977. *Occupations and society: Toward a sociology of the labor market.* New York: John Wiley.

Moore, Kristin A., and Isabel V. Sawhill. 1976. Implications of women's employment for home and family life. In *Women and the American economy,* ed. Juanita M. Kreps. Englewood Cliffs, Prentice-Hall.

Moore, Maurice J., and Martin O'Connell. 1978. Perspectives on family fertility. *Current Population Reports.* Special studies ser. P-23, no. 70. Washington, D.C.: Bureau of the Census.

Moore, Maurice J., and Carolyn C. Rogers. 1979. Some new measurements of first marriages, 1954 to 1977. Washington, D.C.: Population Division, Bureau of the Census. Paper delivered at the Annual Meeting of the Population Association of America, Philadelphia, April 28.

Moos, R. H. 1974. *Family environment scale* (Form R). Palo Alto, Calif.: Consulting Psychologists Press.

Morgenthaler, Eric. 1979. Sweden offers fathers paid paternity leaves: About 10 percent take them. *Wall Street Journal,* Jan. 30.

Mott, P., F. Mann, Q. McLouglin, and D. Warwick. 1965. *Shift work: The social, psychological, and physical consequences.* Ann Arbor, Mich.: University of Michigan Press.

Mueller, Oscar, and Muriel Cole. 1977. Concept wins converts and federal agency. *Monthly Labor Review* 100: 71–74.

Myers, Vincent. 1977. Toward a synthesis of ethnographic and survey methods. *Human Organization* 36 (3): 244–51.

Myrdal, Alva. 1967. Introduction. In *The changing roles of men and women,* ed. E. Dahlstrom and R. Liljestrom. London: Duckworth.

Myrdal, Alva, and Viola Klein. 1956. *Women's two roles: Home and work.* London: Routledge and Kegan Paul.

Nathan, Robert Stuart. 1977. The scheme that's killing the rat-race blues. *New York Times Magazine,* July 8.

National Academy of Sciences, Advisory Committee on Child Development, National Research Council. 1976. *Toward a national policy for children and families.* Washington, D.C.: The Academy.

National Council for Alternative Work Patterns. 1977. *Resource packet: National Conference on Alternative Work Schedules.* Chicago, Ill.: The Council, March.

National Geological Survey. 1976. Report on flexitime program. Washington, D.C.: The Survey.

Nevill, Dorothy, and Sandra Damico. 1974. Development of a role conflict questionnaire for women: Some preliminary findings. *Journal of Consulting and Clinical Psychology* 42: 743.

Newbrough, J. R., et al. 1978. *Families and family institution transactions in child develoment.* Mimeograph. An analysis of the Family Research Program of HEW's Administration for Children, Youth, and Families; Center for Community Studies, George Peabody College for Teachers; and Center for the Study of Families and Children, Institute for Public Policy Studies, Vanderbilt University, April 1.

Nickols, Sharon Y. 1976. Work and housework: Family roles in productive activity. Paper delivered at the Annual Meeting of the National Council on Family Relations, New York City, Oct.

Nie, Norman H., C. Hadlai Hull, Jean G. Jenkins, Karin Steinbrenner, and Dale H. Bent. 1970. *Statistical package for the social sciences.* 2d ed. New York: McGraw-Hill.

Nollen, Stanley. 1979. *New patterns of work.* Scarsdale, N.Y.: Work in America Institute.

Nollen, Stanley, and Virginia Hilder Martin. 1978. *Alternative work schedules.* Pt. 1. *Flexitime.* New York: AMCOM, a division of American Managements Associates.

Nollen, Stanley, Brenda Eddy, and Virginia Hilder Martin. 1978. *Part-time employment, the manager's perspective: An exploratory analysis of employer level issues.* New York: Praeger.

Nordheimer, Jon. 1977. The family in transition: A challenge from within. *The New York Times,* Nov. 17.

Norwood, Janet L. 1977. New approaches to statistics on the family. *Monthly Labor Review* 100 (7): 31–34.

Nunnally, Jum C. 1967. *Psychometric theory.* New York: McGraw-Hill.

Nye, F. Ivan. 1976. *Role structure and analysis of the family.* Beverly Hills, Calif.: Sage Publications.

Nye, F. Ivan and F. Berardo. 1966. *Emerging conceptual frameworks in family analysis.* New York: MacMillan.

Oakley, Ann. 1976. *Woman's work.* New York: Vintage.

Office of Personnel Management. 1969. *Federal Personnel Manual,* bk. 630, supp. 990-2, subchap. 53, Annual Leave, p. 630-17. Rev. ed.; Washington, D.C.: The Office, July; orig. pub. Oct. 16, 1964.

Oppenheimer, Valerie Kincade. 1974. The life-cycle squeeze: The interaction of men's occupational and family life cycles. *Demography* 11: 227–45.

Orden, Susan R., and Norman M. Bradburn. 1969. Working wives and marriage happiness. *American Journal of Sociology* 74: 392–407.

Ory, Marcia G., and Robert K. Leik. 1978. *Policy and the American family: A manual for family impact analysis* Report no. 14, Minnesota Family Study Center. Minneapolis: University of Minnesota.

O'Toole, James. 1974. *Work and the quality of life: Resource papers for work in America.* Cambridge, Mass.: MIT Press.

Owen, John D. 1975. Testimony. In U.S., Congress, House of Representatives, Committee on Post Office and Civil Service, Subcommittee on Manpower and Civil Service Hearings, *Alternate work schedules and part-time career opportunities in the federal government* (serial no. 94-53), p. 58.

______. 1977a. Flexitime: Some management and labor problems of the new flexible hour scheduling practices. *Industrial and Labor Relations Review* 30: 152–61.

______. 1977b. An empirical analysis of the voluntary part-time labor market. Draft report (Research and Development Contract), prepared for the Manpower Administration, U.S. Department of Labor. Aug. 31.

______. 1978a. Hours of work in the long run: Trends, explanation, scenarios, and implications. *Work Time and Employment* 28: 31–64.

______. 1978b. The long-run prospects for alternative work schedules. Paper, Wayne State University.

Paradise, Louis. 1980. Personal communication.

Parker, Beulah. 1975. *A mingled yarn: Chronicle of a troubled family.* New Haven: Yale University Press.

Parsons, Talcott. 1947. *The structure of social actions.* Glencoe, Ill.: Free Press.

______. 1959. The social structure of the family. In *The family: Its function and destiny,* ed. R. Anshen. Rev. ed., New York: Harper.

______. 1977. The changing economy of the family. Mimeograph, for the Family Economic Behavior Seminar, American Council of Life Insurance, Washington, D.C., Nov. 16.

Parsons, Talcott, and R. Bales. 1955. *Family socialization and interaction process.* Glencoe, Ill.: Free Press.

Pearlin, Leonard I. 1975. Sex roles and depression. In *Life-span developmental psychology: Normative life crises*, ed. Nancy Dalton. New York: Academic Press.

Pearlin, Leonard I., and Joyce S. Johnson. 1977. Marital status, life strains, and depression. *American Sociological Review* 42: 704–15.

Pearlin, Leonard I., and Carmi Schooler. 1978. The structure of coping. *Journal of Health and Social Behavior* 19: 2–21.

Personnel Journal. 1975. Flexitime—A social phenomenon, (June), pp. 318–19.

Piotrkowski, Chaya. 1979a. Comments at theory construction workshop, National Council on Family Relations, Annual Meeting, Boston, Aug.

______. 1979b. *Jobs, families and everyday life*. New York: Basic Books.

Pleck, Elizabeth. 1976. Two worlds in one: Work and family. Mimeograph. University of Michigan, Sept.

Pleck, Joseph H. 1976. Men's new roles in the family: Housework and child care. Paper delivered Nov. 10–12, 1975, at the Ford Foundation/Merrill-Palmer Institute Conference on the Family and Sex Roles, Center for the Family, University of Massachusetts, Amherst, Mass. Rev., Dec.

______.1977a. Developmental stages in men's lives: How do they differ from women's? Paper delivered at the Resocialization of Sex Roles: Challenge for the 1970s, a conference sponsored by the Commission on the Occupational Status of Women, National Vocational Guidance Association, Waldenwoods Conference Center, Hartland, Mich., July.

______. 1977b. The work-family role system. *Social Problems* 24: 417–27.

______. 1978a. Women: Work and personality in the middle years. Memorandum to SSRC Study Groups, University of Massachusetts, Amherst, Mass., March 15.

______. 1978b. Wives' employment, role demands, and adjustment. Grant application proposal to Public Health Services, HEW, June 23.

______. 1979. Men's family work: Three perspectives and some new data. *The Family Coordinator* 28: 481–88.

Pleck, Joseph H., and J. Sawyer, eds. 1974. *Men and masculinity*. Englewood Cliffs, N.J.: Prentice-Hall/Spectrum.

Pleck, Joseph H., Graham L. Staines, and Linda Lang. 1978. Work and family life: First reports on work-family interference and workers' formal childcare arrangements, from the 1977 Quality of Employment Survey. Paper prepared under contract with U.S. Department of Labor. Rev., Sept.

Pless, I. B., and B. Satterwhite. 1973. A measure of family functioning and its application. *Social Science and Medicine* 7: 613–21.

Polatnik, M. 1973. Why men don't rear children: A power analysis. *Berkeley Journal of Sociology* 18: 458–61.

Poloma, Margaret M., and T. Neal Garland. 1971. The married professional woman: A study in the tolerance of domestication. *Journal of Marriage and the Family* 33: 531–40.

Poor, Riva, ed. 1973. *4 days, 40 hours, and other forms of the rearranged workweek.* New York: Mentor New American Library.

Portner, Joyce. 1978. Flexitime: Reported effects on families. Paper delivered at the National Council on Family Relations, Philadelphia, Oct.

Poston, Ersa. 1978. Testimony. In U.S., Congress, Senate, Committee on Governmental Affairs, *Flexitime and part-time legislation* (serial no. .32-422 O), pp. 20–60.

Powell, Barbara, and Marvin Reznikoff. 1976. Role conflict and symptoms of psychological distress in college-educated women. *Journal of Consulting and Clinical Psychology* 44: 173–479.

Presser, Harriet B. 1977. Female employment and the division of labor within the home: A longitudinal perspective. Paper delivered at the Annual Meetings of the Population Association of America, St. Louis, Mo., April.

______. 1978. Personal communication. Department of Sociology, University of Maryland.

Quinn, Robert P., and Linda J. Shepard. 1974. *The 1972–73 quality of employment survey, descriptive statistics, with comparison data from the 1969–70 survey of working conditions.* Report to the Employment Standards Administration, U.S. Department of Labor. Survey Research Center, Ann Arbor, Mich.

Racki, George H. E. M. 1975. The effects of flexible working hours. Ph.D. diss., University of Lausanne.

Rahe, R. H. 1972. Subjects' recent life changes and their near-future illness reports. *Annals of Clinical Research* 4: 250–65.

______. 1974. The pathway between subjects' recent life changes and their near-future illness reports: Representative results and methodological issues. In *Stressful life events*, ed. B. S. Dohrenwend and B. P. Dohrenwend. New York: John Wiley.

Rainwater, Lee. 1970. *Behind ghetto walls*. Chicago: Aldine.

Rallings, E. M., and F. Ivan Nye. 1979. Wife-mother employment, family, and society. *Contemporary Theories About the Family* 1: 203–26.

Rapoport, Rhona, and Robert N. Rapoport. 1976. *Dual career families reexamined*. New York: Harper and Row.

Rapoport, Rhona, Robert N. Rapoport, Ziona Strelitz, and Stephen Kew. 1977. *Fathers, mothers, and society: Towards new alliances*. New York: Basic Books.

Rapoport, Rhona, Robert N. Rapoport, and Victor Thiessen. 1974. Couple symmetry and enjoyment. *Journal of Marriage and the Family* 36: 588–91.

Rapoport, Robert N., and Rhona Rapoport. 1965. Work and family in contemporary society. *American Sociological Review* 30: 381–94.

———. 1971. *Dual career families*. London: Penguin.

———, eds. 1978. *Working couples*. New York: Harper and Row.

Raymond, J. Gene. 1978. Testimony. In U.S., Congress, Senate, Committee on Governmental Affairs, *Flexitime and part-time legislation* (serial no. .32-422 O), p. 63.

Reinhold, Robert. 1977. The trend toward sexual equality: Depth of transformation uncertain. *The New York Times*, Nov. 30.

Renteria, Dorothy. 1977. Testimony. In U.S., Congress, House of Representatives, Committee on Post Office and Civil Service, Subcommittee on Employee Ethics and Utilization Hearings, *Part-time employment and flexible work hours* (serial no. 95-28), p. 131.

Rich, Adrienne. 1976. *Of woman born: Motherhood as experience and institution*. New York: Norton.

Rist, Ray C. 1975. Ethnographic techniques and the study of an urban school. *Urban Education* 10: 86–108.

Roberts, Elizabeth J., and John Gagnon. 1978. "Family life and sexual learning: A study of the role of parents in the sexual learning of children," vol. 1. *A summary report*. Mimeograph. Project on Human Sexual Development, Population Education, Inc., Cambridge, Mass.

Roberts, Elizabeth J., David Kline, and John Gagnon. 1978. "Family life and sexual learning: A study of the role of parents in the sexual learning of children," vol. 3. Mimeograph. Cambridge, Mass.: Project on Human Sexual Development, Population Education, Inc.

Robinson, John P. 1977a. *How Americans use time: A social-psychological analysis.* New York: Praeger.

______. 1977b. *How Americans use time in 1965.* Ann Arbor, Mich.: Institute for Social Research.

Robinson, John P., Philip E. Converse, and Alexander Szalai. 1972. Everyday life in twelve countries. In *The use of time: Daily activities of urban and suburban populations of twelve countries,* ed. A. Szalai. Netherlands: Mouton.

Roiphe, Ann. 1968. *Up the sandbox.* New York: Simon and Schuster.

Rosenberg, Charles, ed. 1975. *The family in history.* Philadelphia: University of Pennsylvania Press.

Rosenberg, Morris. 1968. *The logic of survey analysis.* New York: Basic Books.

Ross, Heather L., and Isabel V. Sawhill. 1977. The family as economic unit. *The Wilson Quarterly* 1: 84–88.

Rossi, Alice S. 1977. A biological perspective on parenting. *Daedalus* 106.

Rothman, Sheila M. 1978. *Woman's proper place.* New York: Basic Books.

Rubin, Lillian. 1976. *Worlds of pain: Life in the working-class family.* New York: Basic Books.

Safilios-Rothschild, Constantina. 1970. The influence of the wife's degree of work commitment on some aspects of family organizations and dynamics. *Journal of Marriage and the Family* 32: 681–91.

______. 1974. *Women and social policy.* Englewood Cliffs, N.J.: Prentice-Hall.

______. 1976. Dual linkages between the occupational and family systems: A macrosociological analysis. *Signs* 1: 51–60.

Sanders, William B. 1974. *The sociologist as detective: An introduction to research methods.* New York: Praeger.

Sarbin, Theodore R., and Vernon L. Allen. 1968. Role theory. In *The handbook of social psychology,* ed. Gardner Lindzey and Elliot Aronson. Reading, Mass.: Addison-Wesley.

Sawhill, Isabel V. 1974. Perspectives on women and work in America. In *Work and the quality of life,* ed. James O'Toole. Cambridge, Mass.: MIT Press.

______. 1977. Economic perspectives on the family. *Daedalus* 106: 115–25.

Scanzoni, John. 1979. Strategies for changing male family roles: Research and practice implications. *The Family Coordinator* 28: 435–42.

Schein, Virginia E., Elizabeth H. Maurer, and Jan F. Novak. 1977. Impact of flexible working hours on productivity. *Journal of Applied Psychology* 62: 463–65.

Schopp-Schilling, Hanna-Beate. 1977. The changing roles of women and men in the family and in society. Report on an International Conference, Aspen Institute, Berlin, Germany, Feb.–March.

Schroeder, Patricia. 1975. Testimony. In U.S., Congress, House of Representatives, Committee on Post Office and Civil Service, Subcommittee on Manpower and Civil Service Hearings, *Alternate work schedules and part-time career opportunities in the federal government* (serial no. 94-53), pp. 20–23.

______. 1978. Testimony. In U.S., Congress, Senate, Committee on Governmental Affairs, *Flexitime and part-time legislation* (serial no. .32-422 0), pp. 16–20.

Scott, Joan and Louise Tilly. 1975. Women's work and the family in nineteenth-century Europe. In *The family in history*, ed. C. Rosenberg. Philadelphia: University of Pennsylvania Press.

Scott, Robert, and Alan Howard. 1970. Models of stress. In *Social stress*, ed. Sol Levine and Norman A. Scotch. Chicago: Aldine Publishing Co.

Seidenberg, R. 1975. *Corporate wives—corporate casualties.* New York: Anchor.

Selye, H. 1956. *The stress of life.* New York: McGraw-Hill.

Sexton, Patricia Cayo. 1977. *Women and work* (R and D Monograph 46). Washington, D.C.: Employment and Training Administration, U.S. Department of Labor.

Shinn, Marybeth. 1978. Father absence and children's cognitive development. *Psychology Bulletin* 85 (2): 295–324.

Shuck, Stuart. 1977. Testimony. In U.S., Congress, House of Representatives, Committee on Post Office and Civil Service, Subcommittee on Employee Ethics and Utilization Hearings, *Part-time employment and flexible work hours* (serial no. 95-28), pp. 154–55.

Siegman, Aron Wolfe. 1966. Father absence during early childhood and antisocial behavior. *Journal of Abnormal Psychology* 71: 71–74.

Silverstein, Pam, and Jozetta H. Srb. 1979. Flexitime: Where, when, and how? *Key Issues Series* 24, New York State School of Industrial and Labor Relations Newsletter, Cornell University.

Sinclair, Ward. 1978. Sweeping reorganizations proposed: Commerce—a $3 billion jumble. *The Washington Post*, Dec. 18.

Skolnick, Arlene. 1974. *The intimate environment: Exploring marriage and the family.* Boston: Little, Brown.

______. 1975. The family revisited: Themes in recent social science research. *Journal of Interdisciplinary History* 5: 703–19.

Skolnick, Arlene, and Jerome H. Skolnick. 1974. *Intimacy, family, and society.* Boston: Little, Brown.

Smith, Lee. 1977. Flexitime: A new work style catches on. *Dunn's Review*, March, 109 (3): 61–64.

Smith, Ralph E. 1977a. The effects of hours rigidity on the labor market status of women. Mimeographed remarks at the National Conference on Alternative Work Schedules, Chicago, Ill., March.

______. 1977b. Letter. In U.S., Congress, House of Representatives, Committee on Post Office and Civil Service, Subcommittee on Employee Ethics and Utilization Hearings, *Part-time employment and flexible work hours* (serial no. 95-28), p. 241.

______. 1977c. *Projecting the size of the female labor force: What makes a difference?* Washington, D.C.: The Urban Institute.

______. 1977d. Sources of growth of the female labor force 1971–75. *Monthly Labor Review* 100 (8): 27–28.

Smuts, Robert W. 1959. *Women and work in America.* New York: Columbia University Press.

Social Security Administration. 1974. Do-it-yourself work schedule. *Oasis* (July).

Sokolwoska, M. 1965. Some reflections on the different attitudes of men and women towards work. *International Labor Review* 92: 35–50.

Solarz, Hon. Stephen J. 1977. Testimony. In U.S., Congress, House of Representatives, Committee on Post Office and Civil Service, Subcommittee on Employee Ethics and Utilization Hearings, *Part-time employment and flexible work hours* (serial no. 95-28), pp. 5–110.

Spain, Jayne B. 1975. Testimony. In U.S., Congress, House, Committee on Post Office and Civil Service, Subcommittee on Manpower and Civil Service Hearings, *Alternate work schedules and part-time career opportunities in the federal government* (serial no. 94-53), p. 28.

Spanier, Graham B. 1976. Measuring dyadic adjustment: New scales for assessing the quality of marriage and similar dyads. *Journal of Marriage and the Family* 38: 15–28.

Spellman, Gladys N. 1977. Testimony. In U.S., Congress, House of Representatives, Committee on Post Office and Civil Service, Subcommittee on Employee Ethics and Utilization, *Part-time employment and flexible hours* (serial no. 95-28), pp. 1–5.

Spiegel, J. P. 1957. Resolution of role conflict in the family. *Psychiatry* 20: 1–16.

Spock, B. 1946. *Baby and child care.* New York: Pocket Books.

______. 1974. *Raising children in a difficult time.* New York: Norton.

Stack, Carol B. 1974. *All our kin: Strategies for survival in two black families.* New York: Harper and Row.

Staines, Graham L. 1978. Personal communication. Survey Research Center, University of Michigan.

Staines, Graham L., and Robert P. Quinn. 1979. American workers evaluate the quality of their jobs. *Monthly Labor Review* 102: 3–12.

Staines, Graham L., et al. 1978. Wives' employment status and mental adjustment: Yet another look. *Psychology of Women Quarterly* 5.

Stein, Barry, Allan R. Cohen, and Herman Gadon. 1976. Flexitime: Work when you want to. *Psychology Today* 10 (1): 40–43; 80.

Steiner, Gilbert Y. 1977. *Early childhood and family policies.* Position paper for staff and trustees of the German Marshall Fund. Washington, D.C.: the Brookings Institution.

Stone, Lawrence. 1975. The rise of the nuclear family in early modern England. In *The family in history,* C. Rosenberg. ed. Philadelphia: University of Pennsylvania Press.

Strauss, Murray A. 1969. *Family measurement techniques.* Minneapolis: University of Minnesota Press.

Strauss, Murray A., and Bruce W. Brown. 1978. *Family measurement techniques: Abstracts of published instruments, 1935–1974.* Rev. ed. Minneapolis: University of Minnesota Press.

Stress and Families Project. 1979. Lives in stress: A context for depression. Rev. mimeograph, Oct.

Strober, Myra H. 1977. Women and men in the world of work: Present and future. In *Women and men: Changing roles, relationships and perceptions.* ed. Libby A. Cater, Anne Firor Scott, and Wendy Martyn. New York: Praeger, 1977.

Strong, E. K., Jr. 1943. *Vocational interests of men and women.* Stanford, Calif.: Stanford University Press.

Sudman, Seymour, and Norman M. Bradburn. 1974. *Response effects in surveys: A review and synthesis*. Chicago: Aldine.

Sullivan, Anne. 1978. Statement. In U.S., Congress, Senate, Committee on Governmental Affairs, *Flexitime and part-time legislation* (serial no. .32-422 O), pp. 277–85.

Super, D. E., and J. O. Crites. 1962. *Appraising vocational fitness*. Rev. ed. New York: Harper and Row.

Survey Research Center. 1977a. Men's two roles: Work and family. Questionnaire for the Institute for Social Research, University of Michigan, Ann Arbor, Mich., July.

______. 1977b. The 1977 quality of employment survey. Questionnaire for the Institute for Social Research, University of Michigan, Ann Arbor, Mich., Fall.

Swart, J. Carroll. 1978. *A flexible approach to working hours*. New York: AMCOM, a division of American Management Associates.

Sykes, Richard E. 1978. Toward a theory of observer effect in systematic field observation. *Human Organization* 37: 148–56.

Szalai, Alexander, ed. 1972. *The use of time: Daily activities of urban and suburban populations of twelve countries*. Netherlands: Mouton.

Taeuber, Karl E., and James A. Sweet. 1976. Family and work: The social life cycle of women. In *Women and the American economy*, ed. Juanita M. Kreps. Englewood Cliffs, N.J.: Prentice-Hall.

Tognoli, Jerome. 1979. The flight from domestic space: Men's roles in the household. *The Family Coordinator* 28: 599–607.

Troll, Lillian E. 1969. Issues in the study of the family. *Merrill-Palmer Quarterly* 15: 221–26.

Tucker, M. J. 1974. The child as beginning and end: Fifteenth- and sixteenth-century English childhood. In *The history of childhood*, ed. Lloyd deMausse. New York: Harper and Row.

Turner, Ralph H. 1970. *Family interaction*. New York: John Wiley.

U.S. Civil Service Commission, Bureau of Programs and Standards. 1962. *A history of hours of work in the federal service*. Washington, D.C.: G.P.O.

U.S. Civil Service Commission, Bureau of Personnel Management Information Systems. 1976. *Federal civilian workforce statistics: Occupations of federal white-collar workers* (SM 56-12). Washington, D.C.: G.P.O.

U.S. Comptroller General. 1977. *Benefits from flexible work schedules—legal limitations remain* (report to the Congress; EPCD—78-62).

Washington, D.C.: General Accounting Office.

U.S., Congress, House of Representatives. 1977a. A Bill to authorize federal agencies to experiment with flexible and compressed employee work schedules. Ninety-fifth Cong., 1st sess., H.R. 7814, June 15.

______. 1977b. A Bill to amend Title 5, U.S. Code, to establish a program to increase part-time career employment within the civil service. Ninety-fifth Cong., 1st sess., H.R. 10126, Nov. 29.

U.S., Congress, House of Representatives, Committee on Post Office and Civil Service, Subcommittee on Manpower and Civil Service. 1975. *Alternate work schedules and part-time career opportunities in the federal government; Hearings on H.R. 6350, H.R. 9043, H.R. 3925, and S. 792* (serial no. 94-53). Ninety-fourth Cong., 1st sess., Sept. 29–30; Oct. 7.

U.S., Congress, House of Representatives, Committee on Post Office and Civil Service. 1976. *Experiments to test flexible and compressed work schedules for federal employees* (*report no. 94-1033*). Ninety-fourth Cong., 2d sess., April 12.

______. 1978. *Federal employees flexible and compressed work schedules act of 1978: Report to accompany H.R. 7814* (report no. 95-912). Ninety-fifth Cong., 2d sess., Feb. 28.

U.S., Congress, House of Representatives, Committee on Post Office and Civil Service, Subcommittee on Employment Ethics and the Utilization of the Committee on Post Office and Civil Service. 1977. *Part-time employment and flexible work hours: Hearings on H.R. 1627, H.R. 2732, and H.R. 2930* (serial no. 95-28). Ninety-fifth Cong., 1st sess., May 24, June 29, July 8, and Oct. 4.

U.S., Congress, Senate, Committee on Governmental Affairs. 1978. *Flexitime and part-time legislation: Hearings on S. 517, S. 518, H.R. 7814, and H.R. 1026*. Ninety-fifth Cong., 2d sess., June 29.

U.S., Congress, Senate, Committee on Post Office and Civil Service. 1974a. *Flexible hours in the federal service* (report no. 93-1143). Ninety-third Cong., 2d sess., Sept. 11.

______. 1974b. *Flexible hours employment: Hearings on S. 2022* (serial no. .32-422 O). Ninety-third Cong., 2d sess., Sept. 26.

U.S. Department of Commerce. 1975. The U.S. Department of Commerce building: Features of interest. Washington, D.C.: Department of Commerce.

VanderVeen, Ferdinand. N.d. Content dimensions of the family concept test and their relation to childhood disturbance. Chicago: Institute for Juvenile Research.

Vanek, Joann. 1974. Time spent in housework. *Scientific American* Nov. 231: 116–120.

______. 1977. The new family equality: Myth or reality? Paper delivered at the annual meetings of the American Sociological Association, Chicago, Ill. Sept.

Viorst, Judith. 1968. *It's hard to be hip over thirty, and other tragedies of married life.* Mountain View, Calif.: World.

Waite, L. 1976. Working wives, 1940–60. *American Sociological Review* 4: 65–80.

Waldman, Elizabeth, Allyson Sherman Grossman, Howard Hayghe, and Beverly L. Johnson. 1979. Working mothers in the 1970s: A look at the statistics. *Monthly Labor Review* 102: 39–49.

Walker, Kathryn E., and Margaret E. Woods. 1976. *Time use: A measure of household productions of family goods and services.* Washington, D.C.: Center for the Family, The American Home Economics Association.

Warwick, Donald P., and Charles A. Lininger. 1975. *The sample survey: Theory and practice.* New York: McGraw-Hill.

Weiss, Robert S. 1968. Issues in holistic research. In *Institutions and the person: Papers presented to Everett C. Hughes*, ed. Howard S. Becker, Blanche Geer, David Riesman, and Robert S. Weiss. Chicago: Aldine.

Weiss, Robert S., and M. Rein. 1970. The evaluation of Broad-aim Programs: Experimental design and its difficulties and an alternative. *Administrative Science Quarterly* 15: 94–109.

Weller, R. 1968. The employment of wives, dominance, and fertility. *Journal of Marriage and the Family* 30 (3): 437–42.

Wells, Myra. N.d. Flexitime status report. Mimeograph, Maritime Administration, U.S. Department of Commerce, Washington, D.C.

Westinghouse Learning Corporation. 1969. *The impact of Head Start: An evaluation of the Head Start experience on children's cognitive and affective development.* Ohio: Westinghouse Learning Corporation, Ohio University.

Whiting, Beatrice Blyth, and John W. M. Whiting. 1975. *Children of six cultures: A psycho-cultural analysis.* Cambridge, Mass.: Harvard University Press.

Whittaker, William G. 1978. *Alternative work schedules and part-time career opportunities in the federal government: A legislative review.* Washington, D.C.: Congressional Research Service.

Wilensky, Harold L. 1961. The uneven distribution of leisure: The impact of economic growth on "free time." *Social Problems* 9: 32–56.

Wilson, L., and G. McDonald. 1977. Family impact analysis and the family policy advocate: The process of analysis (Family Impact Ser. no. 4). Minneapolis: Minnesota Family Study Center, University of Minnesota.

Winett, Richard A. 1978. Personal conversations regarding his study of effects of flexitime on family life in two federal agencies. Institute for Behavioral Research, Silver Spring, Md.

Winett, Richard A., and Michael S. Neale. 1978. Family life and the world of work: A preliminary report on the effects of flexitime. Paper from the Institute for Behavioral Research, delivered at the meeting of the American Psychological Association, Toronto, Aug.

Winnicott, D. W. 1963/74. *The child, the family, and the outside world.* Harmondsworth, England: Penguin.

Wiseman, Jacqueline P. 1974. The research web. *Urban life and culture* 3: 317–28.

Wispé, Lauren G. 1955. A sociometric analysis of conflicting role-expectancies. *American Journal of Sociology* 61: 134–37.

Wortis, R. P. 1974. The parental mystique. In *Intimacy, family, society,* ed. A. Skolnick and J. H. Skolnick. Boston: Little, Brown.

Wright, James. 1978. Are working women really more satisfied? Evidence from recent national surveys. *Journal of Marriage and the Family* 40: 301–14.

Wylie, Lawrence. 1958. *Village in the Vaucluse.* Cambridge, Mass.: Harvard University Press.

Young, Michael, and Peter Willmott. 1973. *The symmetrical family.* New York: Pantheon.

Zablocki, B. 1976. The use of crisis as a mechanism of social control. In *Social change: Explorations, diagnoses, and conjectures,* ed. G. K. Zollschan and W. Hirsch. Cambridge, Mass.: Schenkman.

Zagoria, Sam. 1974. Flextime—a city employee pleaser. *Nation's Cities* 12: 42–46.

Zaretsky, Eli. 1973. *Capitalism, the family, and personal life.* New York: Harper and Row.

Zetterberg, H. L. 1965. *On theory and verification in sociology*. 3d ed. Totowa, N.J.: Bedminster Press.

Zinsser, Caroline. 1978. Position paper, Workshop on the Impact of Corporate Policies on the Family, Aspen Institute for Humanistic Studies, in Aspen, Colo., Aug.

INDEX